Schumacher Briefing

THE NATURAL STEP

Towards A Sustainable Society

David Cook

Published by Green Books
for The Schumacher Society

First published in 2004
by Green Books Ltd
Foxhole, Dartington,
Totnes, Devon TQ9 6EB
www.greenbooks.co.uk
info@greenbooks.co.uk

for The Schumacher Society
The CREATE Centre, Smeaton Road,
Bristol BS1 6XN
www.schumacher.org.uk
admin@schumacher.org.uk

Cover design by Rick Lawrence
samskara@onetel.com

Printed by J. W. Arrowsmith Ltd, Bristol, UK
Text printed on Corona Offset 100% recycled paper
Covers are 80% recycled material

ISBN 1 903998 47 6

THE NATURAL STEP

David Cook's connection to sustainable development began with the development of The Natural Step (TNS) in the UK.

His previous career involved various senior management appointments in the public sector, including work with schools and colleges, social services, the probation service and other local government activities. David spent three years as Chief Executive of a public/private Project Company, to plan, fund and deliver a new University in the UK. The project won a national award for a public sector and a private partnership development. David has always had particular interest in the human behavioural aspects of organisational change, and in overcoming barriers to creativity and development.

Working with leading figures in the environment and sustainability movement, David was responsible for establishing TNS in the UK. A primary focus was on working with big corporations to challenge them with the realities of sustainable development, which has led to the successful TNS Pathfinder Programme.

With the aim of expanding the application of The Natural Step Framework to new sectors, David has worked with senior managers and operational teams in many organisations around the world. He has played a leading part in the expansion of TNS's approach to the social dimension of sustainability and Corporate Social Responsibility. David was appointed as Chief Executive of TNS International from April 2004.

Contents

Foreword

It is good timing for The Natural Step (TNS) to collaborate with the Schumacher Society UK in this publication. E. F. Schumacher himself was one of the first people to talk about sustainable development. He set out the problems of a world in which global economics becomes too distant from individuals and communities. He saw the need for a thorough rethinking of the values and the mindset that drives the post-industrial age. Sustainable development as a concept has made real progress in recent decades, but its application is still too limited to deal with the scale of the sustainability challenge.

The Natural Step, since its inception 15 years ago, has developed a framework aimed precisely at promoting more profound and systematic application of sustainable development around the world. This book deals with the theory and practice of that framework. Though applying a framework is at least as much about passion, inspiration and artistry as rigour, the development of the framework itself has been based upon scientific principles and peer review. It tells us how to achieve success within the boundaries of the system to which we belong. The framework is both comprehensive and comprehensible. In other words it deals with everything, but can be understood without technical knowledge. So we have built up a language and a way of seeing the world that can help decision-makers to put sustainable development into action. Many obstacles remain, and two of those occupy us the most right now.

Firstly, we need more finance to fuel the dissemination of our work and our message. TNS is now embarked upon a new step in its history to become a truly international organisation. Our mission is the dissemination of the framework to all decision-makers, and at all levels, in all parts of the world. For that we need more resources.

The second obstacle can be summarised as the seduction of comforting and familiar habits. Change of thinking and behaviour does not happen easily. Our familiar habits are reinforced by the ways of the world in which we live. Through the media, marketing, and the market itself, we are too easily constrained to reach for new ways of thinking. In business and in policy making it is easiest to stick with what we know. It is comforting to feel in control by working with the detail of our own backyard. Science all too often follows a similar reductionist pattern and does not help us to see the wider picture or the wider consequences of our actions. The difficulty for those in science and in policy who would attempt to broaden our horizon, is that they are in danger of being accused of prescribing the way our lives should be.

Because of the underpinning science of The Natural Step Framework we avoid the charge of being prescriptive; we simply communicate the way things are. Paradoxical as it may seem, we do that in a manner that highlights the opportunities through being thorough in the description of constraints. By using the framework, those who engage with TNS are able to open up a diverse range of styles for playing the sustainability game. We do not displace the many excellent tools and techniques that are already out there (EMAS, Eco-Footprinting, Factor X, Life Cycle Analysis, Zero Emissions, Natural Capitalism, etc). On the contrary, we set out to improve on the applicability of such good tools by highlighting them from the perspective given by the TNS Framework.

The only way to overcome the difficulties of apathy and resistance in terms of progress towards a more sustainable society is to carry on getting the message out as broadly as we can. This Briefing meets that purpose. It also gives us the chance to update readers with some of the more recent framework developments and success stories. Above all our ambition is that, in a world that seems to be ever more cynical and partisan, you will find in this Schumacher Briefing reasons to be hopeful, and the means to make a difference.

It would be inappropriate, and probably even counterproductive, to here produce any flattering specifics of David Cook, a close friend and colleague of mine. David, let me just say how grateful I am for this book, and all other aspects of your work.

Karl-Henrik Robèrt
Chairman, The Natural Step International

Acknowledgements

Many people have contributed to the story which is set out in this Briefing. My job has largely been to bring it all together, and to set the context. Karl-Henrik Robèrt has been an enthusiastic contributor and guide throughout. Writing this Briefing has coincided with the preparation of material for the new Masters Course (based on The Natural Step) at Blekinge Institute of Technology, a Swedish university. That has enabled me, with assistance from Dave Waldron, the coordinator of that course, to make use of some of that material.

TNS teams from around the world have contributed. Without their dedicated work there would be no story to tell. Regular encouragement from Jonathon Porritt and Hugh Pidgeon in the UK for me to persist in the exploration of System Condition 4 has been essential. Finally there would have been no prospect of this Briefing being completed and published without the tremendous support and hard work of Marion Wells, and without commissioning, guidance and editing from Herbie Girardet of the Schumacher Society.

Introduction

How do we make economic progress, giving everyone the chance of a fulfilling life, without continuing to damage the natural systems upon which we all depend? Today the challenge of sustainability seems just as urgent as it did fifteen years ago when The Natural Step began. There has been progress, but there is still a long way to go—for sustainable development and for TNS.

The special contribution of The Natural Step is that it brings a science-based understanding to the complex issues we face. In order to build a sustainable society, we first need to know what an un-sustainable society looks like. Science is the best place to look for that knowledge. The problem is that science so often fails to give us a coherent message. For new solutions we do not need more science at the detailed level, simply telling us about the impacts, *downstream,* of our unsustainable behaviour. What we need is a better view of the whole system. We need knowledge about what is going on *upstream,* where the problems begin. We need to see more clearly how our actions affect the whole system. That kind of knowledge needs to inform decisions, policies and strategies at global, national and local levels. We will only start to see progress when leaders in all sectors start to make decisions that relate actions to consequences. It is simply not good enough to think only of our own interests in politics or in commerce. The interdependencies of a more crowded and connected world make it even more vital for us to take a wider systems view.

Changing the mindset to one that is more attuned to thinking about consequences is not easy. Sustainable development is not an easy option, and yet it must become a priority for us all. The opportunity to set out an up-to-date description of how The Natural Step is addressing that priority was one which I was happy to take. The format of the Schumacher Briefing series is ideal for these purposes. An explanation that you can absorb on a train journey is an attractive idea. The Natural

Step tells a story, and that story is the way this Briefing has been put together. It begins with absorbing the scale of the ecological disaster facing us, and the damage we are inflicting upon the systems and cycles of nature. From that awareness we can construct the basic preconditions for a sustainable society—the System Conditions. The Natural Step enables us to imagine what a sustainable society would be like. Only then can we see more clearly the obstacles we have to overcome in order to make that vision a reality. The content of the TNS Framework is outlined in this Briefing.

The ideas that TNS teams have developed with regard to the social aspects of sustainable development are given their own chapter. Recent years have seen a growing interest in the social impacts of decisions made by organisations. For the first time we have set out our evolving approach to the social dimension, in line with the scientific and consensus traditions of TNS. The Natural Step is a unified approach to the complexity of *sustainability*: we see it as a whole system issue that is not divisible into separate economic, social and environmental boxes.

I have also taken the opportunity to tell some stories about The Natural Step framework being put into practice. There are new stories here, not just from business but also from the public sector.

As we enter the next phase of development for TNS as an organisation, I hope that this Briefing will make a contribution to the wider dissemination of the TNS Framework. The urgency of the situation we face cannot be denied. No longer can we hide behind disagreements amongst scientists over details. The fact that we are systematically degrading the capacity of nature to support us, is evident everywhere. Whether we have gone beyond any specific and measurable limit is not the main issue here. The truth is that we are responsible for a systematic build-up of impacts, which will, unless halted, crash into the limits of the very system upon which we utterly depend. It is time for all decision-makers to take these incontrovertible danger signals into account. Imagine if we could succeed in getting policy-makers around the world to think more clearly together, and to consider the longer- term impacts of their decisions. Imagine if we could encourage them to give real leadership to the big project of sustainable development.

The importance of acting together towards a sustainable future is a message that must be broadcast and received everywhere. I wanted to tell the story that first made me listen more carefully. I hope it does the trick for you too.

David Cook
September 2004

Chapter 1

What is The Natural Step, and how has its framework developed?

The Natural Step is an organisation that is all about promoting deeper understanding and commitment to sustainability, and a wider application of sustainable development. Sustainability refers to our own human society being capable of continuing indefinitely. Development that will move society in that direction is what we call *sustainable development*. The key feature of TNS is its science-based definition of a sustainable society. Without that definition, sustainability is open to too many interpretations and too much confusion. Real progress will only be possible with a shared understanding of what success will look like.

In order to arrive at a definition of success—in this case sustainability—we must know *enough* about the system—the environment around us (the biosphere), human societies, and the interactions and flows of materials between the two. The concept of *sustainability* becomes relevant only when we understand the *un-sustainability* inherent in the current activities of society. In what principal ways are we destroying the biosphere's ability to sustain us? This question is answered by looking *upstream*, where the decisions are made that trigger the thousands of negative impacts occurring *downstream*.

The negative impacts of un-sustainability encountered today can—at a fundamental level—be divided into three separate mechanisms by which humans can destroy the biosphere and its ability to sustain society. We are systematically digging so much stuff out of the Earth that nature can't cope; systematically poisoning the system with polluting chemicals; and systematically burning, covering over, and generally laying waste to the living environment. In addition we live in societies that do not give individuals a chance to lead a decent way of life. If people do not have the opportunity to lead fulfilling lives, or even to meet their basic needs, they are not likely to take care of the environment. These basic unsustainable behaviours are expressed more scientifically in The Natural Step System Conditions:

THE NATURAL STEP SYSTEM CONDITIONS

In the sustainable society, nature is not subject to systematically increasing . . .

1. . . . concentrations of substances extracted from the Earth's crust,
2. . . . concentrations of substances produced by society,
3. . . . degradation by physical means;

and, in that society . . .

4. . . . people are not subject to conditions that systematically undermine their capacity to meet their needs.

The System Conditions tell us what a sustainable society would look like. Sustainable development is all about making progress towards that sustainable society. We need policies, strategies and decisions that take us in a sustainable direction. The Natural Step has developed the TNS Framework as a collection of strategic methods and strategic communications to help organisations make real progress. The System Conditions and the framework are set out in more detail in Chapters 4,5 and 6 of this Briefing.

TNS: The organisation

The Natural Step is an international non-profit organisation that began life some 15 years ago in Sweden, and its international office remains in Stockholm. In addition there are now TNS Teams in the US, UK, Australia, South Africa, New Zealand, Israel, Canada, Japan and Brazil, and activities in many other places including France, Italy and Hungary. Our work covers three main programmes:

- Research and Development—science projects and science-based dialogues on complex issues

- Advisory Services—working with organisations to increase their understanding and application of sustainable development
- Outreach—awareness raising and educational programmes.

The story of the growth of this organisation is also the story of the evolution of the TNS Framework. It all began with the trials and tribulations of one man, Dr Karl-Henrik Robèrt. It is impossible to talk about TNS without telling his story. It is a tale that adds life to the framework itself. In its telling and re-telling, this story has almost attained the status of a mythology. For the full low-down you really need to go to the source and see how Karl-Henrik himself has retold the myth.[1]

Karl-Henrik was pursuing his profession as a medical doctor, an oncologist, studying cells and tumours, at a Stockholm medical Institute. His work and his microscope taught him that there are limits within which each cell must function. Realising that we are all made of cells, "even politicians", as he puts it, led him to question why so little attention seemed to be given to the systemic causes of problems.

Karl-Henrik has a deep personal connection to nature and the countryside, which began with his childhood. As an adult he was troubled by what he saw around him as the trashing of nature. He was equally perplexed by the way the environmental debate, even in Sweden, always seemed to divide society rather than unite it. No one seemed to be paying attention to the root causes, or to spotting the connections. He turned to systems thinking to help his own comprehension and to share with others.

There followed an amazing period of Karl-Henrik developing his ideas, researching what others had done and starting to talk to lots of different people about his concern and his conclusions. He started writing a scientific paper on the subject and sent it to many Swedish scientists to get their reactions. After about 22 different drafts, a consensus was reached on some fundamental principles that make the earth's system function. His conversations expanded from talking to the local shopkeeper, to leaders of industry, stars of Swedish media, senior politi-

---

1. Robèrt, Karl-Henrik, *The Natural Step Story: seeding a quiet revolution*, New Society Publishers, Gabriola Island, 2003

cians and even the King of Sweden. All agreed that he was really onto something. His breakthrough resulted in a TV programme, information packs being sent to every Swedish household, and significant backing from influential sections of Swedish society. The Natural Step was born.

From 1989 onwards the team which gathered around Karl-Henrik went on to work with many of the largest corporations in Sweden. The work done in those early years continues to shine through and inspire. It includes terrific breakthroughs with major companies such as Electrolux, IKEA, McDonalds and others. Sustainable development initiatives were happening in other places too, mainly in Europe and the United States. A momentum was gathering, even a movement, and that wider scene surely played a part in helping TNS to get its foothold.

The spread of The Natural Step began quite quickly. During the early to mid-nineties TNS teams were established in several other countries. Big-hitters in the environmental world, such as Paul Hawken in the US and Jonathon Porritt in the UK, were easily attracted to it. It offered them a unique scientific model for the dissemination of sustainability at a time when that concept was too often beset with woolly thinking and confusion. Initially the Swedish originators licensed others to operate TNS in their own countries. By 1999 a network of such licensed organisations in various parts of the world had grown to the point where Swedish colleagues felt able to hand over stewardship to a new organisation—The Natural Step International.

Every place is different and TNS has deliberately allowed experimentation that seemed appropriate to different cultures. This was particularly true between the US and UK, for example, where very different starting points were chosen. In the US there was an emphasis on fairly wide dissemination through conferences and open learning events. In the past five years this has changed, with an increasing emphasis on application within specific organisations. TNS in the UK, working within the sustainability charity Forum for the Future, chose to start working almost exclusively with large businesses. The UK Pathfinder Programme was a deliberate model for the TNS UK team to learn quickly about the practical application of the framework.

The core TNS message about sustainability is built upon science, and such very fundamental science does not change very much. But the application of that message keeps on evolving. So the work keeps on changing, growing and being sustained by our learning from practical applications. As an organisation, TNS has had constant concern for the way its message is communicated. The scientific principles that underpin the framework have been referred to as 'non-negotiable'. That is a hard message to combine with allowing people to apply those principles in their own way. The science of TNS is its heart and soul. We must manage to communicate whilst safeguarding the rigour and scientific credibility of the message. The Natural Step's mission is to disseminate the framework in order to generate better awareness and application of sustainable development. We can't afford to fail.

"The work of Karl-Henrik Robèrt and his colleagues through The Natural Step process is one of the leading examples in the world today of society-wide learning. Learning based on systems thinking and continued dialogue . . . holds great promise for many of the most intractable societal issues of our time."

—Peter Senge, Centre for Organisational Learning, MIT

Chapter 2

The Challenge of Sustainability

What is all the fuss about, and why is this such an urgent wake-up call?

Many of those who take the trouble to find out about sustainable development—and probably most of those reading this Briefing—are well aware of the seriousness of the human impacts upon the environment. Increasingly large numbers of people see the links between those impacts and our socio-economic progress. Limited space here allows me only to summarise the perilous situation we face, without going into great detail. There has long been a kind of mantra amongst some environmentalists that those who do not see the danger are too likely to be overwhelmed by the scale of the problem. Therefore we must be careful not to 'frighten the horses' lest they become too overwhelmed to do anything about it. There is much common sense in this. You must give people hope. At the same time there really is no way of ignoring the threats resulting from our continuing unsustainable behaviour.

I came to work with TNS through a seemingly obscure route, with no professional connection to the world of sustainability or even the environment. Of course now I see how absolutely un-obscure this route really was. There were three particular experiences that now seem important. In the first place, my earliest memories include me standing quiet as a child, still and alone, in a field by a stream on the edge of the housing estate where I grew in the English Midlands. When I was still and quiet, nature was opened up to me. This happened very often and it left a deep impression on me. Since then I have always had a close connection to the outdoors and to nature. That field and that stream have long since been buried beneath an industrial estate.

The second experience was later in life when I became involved in taking groups of young criminals off to the hills for rock climbing. I realised for the first time both how scarred some people can be by their childhoods, and how absolutely foreign and threatening it can be for them to be close to nature. These were thugs in their late teens and early twenties with some bitter deeds in their records, usually involving violence to themselves as well as others. Many of them had literally never seen a mountain stream before, and they didn't know that they belonged to this sweeter world.

The third part of this short venture into my personal story relates to the point that the situation is too overwhelming. When I was first asked about working with Forum for the Future to get TNS off the ground in the UK, I had almost no notion what it was all about. Then I started to read and the more I read the more I became appalled, almost enraged, that there was all this destruction going on in the name of economic progress. And it was not just the things you hear about in the media, but so much more, and so deadly because if its insidiousness. Why weren't we all up in arms about this? How could the environmentalists still be on the sidelines of policy making, when all about us the fabric of life was being torn to shreds?

The point of this short trilogy is simple. I now believe that we are all capable of seeing and enjoying our individual connection to nature. That insight and joy, which some may call a kind of spirituality, is intrinsic to making us who we are and enabling us to cope with the troubles life will surely send our way. The absence of such insight and joy leads to barbarism and violence. To make a difference in getting more people to share such connectivity means overturning a mindset that has taken deep root within the industrialised pulp-fiction culture of today. That means dealing with people on their own turf, and sharing a logic that is compelling and uses the most modern terms to get everyone involved. To my mind that is exactly what The Natural Step achieves. It tells a story of how the world works. It is a story that everyone can understand, a story we all knew once upon a time. Let us hope it can be *twice* upon a time.

In my view, there is no excuse for not starting the message of TNS with a description of how bad things are. Human impact on the planet has reached a point where it is posing grave threats to our future prosperity and security—to our very survival. As a species we have a very short history, but in terms of the lifetime of the planet we have only just arrived. For most of our short history we have lived quite comfortably within natural limits, depending upon the earth's resources without affecting its ability to renew them. But the earth is now experiencing a build up of toxins in its air, soil and water, at a rate which natural systems cannot absorb. The tidying-up process of nature, perfected by natural systems over billions of years, is being overwhelmed by the way we have lived our lives over the past couple of centuries.

We humans are unique amongst living creatures, in our capacity for both creativity and destruction. We learnt to use the earth's resources very quickly. As a result, we have produced enormous advances in civilisation, health and quality of life. Our ability to make the most of nature's resources and services has also enabled us to multiply at an extraordinary rate. In the last 150 years the human population has grown from 1 billion to 6 billion. Only very recently have we come to realise that we cannot continue indefinitely at such a rate of expansion and impact. Our habits of production and consumption are taking us beyond critical thresholds in the use of the earth's resources. The effects of this—air pollution, climate change, loss of biodiversity, deforestation, soil erosion and water scarcity—are increasingly visible everywhere.

Yet, for all our ingenuity, and in spite of the glorious store-cupboard of natural resources all around us, we have still failed to create a world in which everyone has the opportunity to lead a fulfilling life. The statistics are no less shocking for their familiarity: 1.3 billion people live in absolute poverty; 35,000 die each day of starvation; the richest 20% own 85% of the worlds resources, whilst the poorest 20% own less than 2%. If this is the situation now, what will it be like in 2025, when the human population may be in excess of 8 billion?

Climate change is perhaps the most widely known consequence of our collective behaviour. It has received such massive coverage in the media that it would be difficult to find anyone in the developed world who has not at least heard about it. Yet how many understand that it is just one part of a whole battery of threats to our way of life that we have ourselves unleashed. Once again the statistics are almost overwhelming. The rate of extinction of species is said to be 1000 times greater than the natural or background rate of normal extinction. We are losing a species every twenty minutes. Tropical rainforest is disappearing at similarly alarming rates. In the past 200 years we have used fossil fuels and other materials dug out of the ground that it took the earth millions of years to accumulate. Fish stocks are being over-fished to the point of elimination. Fresh water fish are having their genders changed by substances we dump into their habitats. Polar bears have mercury in their bodies that should not be there, and we humans are accumulating synthetic chemicals in our blood.

Aside from the environmental impacts, is life getting any better? It hardly seems so. The rich are getting richer and the poor are getting poorer. War and terrorism are more prevalent in the world today than 50 years ago. Even in the so-called developed world, improvement in overall quality of life is questionable. Stress levels are high. Gaps between the wealthy few and the poorer majority are growing. Even rapid growth in Gross Domestic Product—that infamous measure of economic well-being—is not improving the quality of life for most people. In the US, for example, poverty is increasing and the health of many people is declining persistently. Communities are fragmenting, participation in democracy is in decline, and levels of trust between people and society are worryingly low.

You do not have to be a sustainability guru to be able to link our mistreatment of the environment to the problems we face in global society. People in poverty, people denied access to their cultural heritage, to education, health and justice, are hardly likely to start taking care of nature. When we pollute and overuse nature's systems, and trash nature's capacity to meet our needs, then we inevitably produce poverty, disease, starvation and ill-health. These constantly lead to social decline, tensions, crime, injustice and violence. Persistent persuasion upon us all to consume ever more products, to measure our happiness according to our spending power, and to dispose of what we buy as quickly as possible, just makes things worse instead of better. So a vicious cycle is perpetuated: environmental mess flows as an unavoidable consequence of social mess, and vice-versa.

Is this just human nature taking its course, the constant ups and downs of economic growth and social breakdown, whilst nature looks on indifferent to this strange paradox it has created? Science tells us otherwise. Our impacts upon the earth, the sheer size of our footprints upon the planet, are now so great that we are a serious danger to ourselves. We are actually damaging the big systems of nature that make our lives possible in the first place. The term viability comes into play here. Often used to describe the economic health of a business, this word has its roots in biology—the ability to sustain life. Our viability as a species is under threat.

There is no consensus amongst scientists over the exact degree of the threat we face. It is true that the earth is a very resilient system. It has withstood many shocks in the past, including meteorites, ice ages and volcanic eruptions. What we do know, however, is that current levels of resource use and pollution are taking us far beyond the earth's restorative capacity, so that we urgently need to look for alternatives. It is not so much a question of using up stocks of materials. It is not the loss of species or the emission of pollutants to the atmosphere per se that is the concern of sustainable development. It is the fact that such impacts are systematically increasing. The real and present danger lies in the rate at which we are overwhelming the systems of nature. We are damaging the systems that deal with our waste in particular, which regulate the atmosphere, ensure useable energy supplies and provide the diversity of natural organisms that are essential to our sophisticated economic progress.

The new emphasis has to be on quality not quantity—development not growth. The difference between economic development and economic growth is profound, and particularly important if one regards economics as reflecting the direction of society as a whole. There are few better ways of understanding this disconnection between growth and development than the writings of economist Herman Daly. He states the following:

"The two processes are distinct—sometimes linked, sometimes not. For example, a child grows and develops simultaneously: a snowball or a cancer grows without developing; the planet earth develops without growing. Economies frequently grow and develop at the same time, but can do either separately. But since the economic is a sub-system of a finite and not growing eco-system, then as growth leads it to incorporate an ever larger fraction of the total system into itself, its behaviour must more and more approximate the behaviour of the total system, which is development without growth. It is precisely the recognition that growth in scale ultimately becomes impossible—and already costs more than it is worth—that gives rise to the urgency of the concept of sustainable development. Sustainable development is development without growth in the scale of the economy beyond some point that is within biospheric carrying capacity." [2]

The first telling phrase here is that planet earth develops without growing. There are limits, systemic limits revealed to us by science, beyond which the system simply will not function: as we push relentlessly up against those system limits, we start to see repercussions.

We are unlikely to be able to overwhelm nature's wonderful and immensely complex systems entirely. But if we make enough of a mess it is most likely that nature will simple chalk us off, like an unwanted dish on the evolutionary menu. We will no longer be 'today's special'. But nature will continue.

Ours is the first generation to have been able to look down on the earth from above, to be able to conceive of it as a system, and to understand its limits. In our awareness, we are unique in nature. We are even able to unravel and interpret the very processes of evolution. The human brain is itself a result of quality development. We are capable of new thinking and new behaviour. We can visit the origins of our deepest emotional responses, and we can explore the workings of the universe. With that capacity, and with such ever-expanding awareness, comes responsibility. Many of the solutions necessary to restore the earth's ecological and social balance already exist. Now we *must* start putting them into practice. That is the challenge of sustainable development.

The Natural Step applies some very fundamental scientific knowledge to that challenge. We use that knowledge in order to tell a story which people can grasp without being overwhelmed. The story raises new understanding that is absolutely essential for real progress.

---

2. Daly, Herman, *Beyond Growth: the economics of sustainable development*, Beacon Press, Boston, 1996.

Chapter 3

Understanding the Basics

How can science help us understand what is going on? What can we learn about our habitat from science? What will that tell us about our economic system?

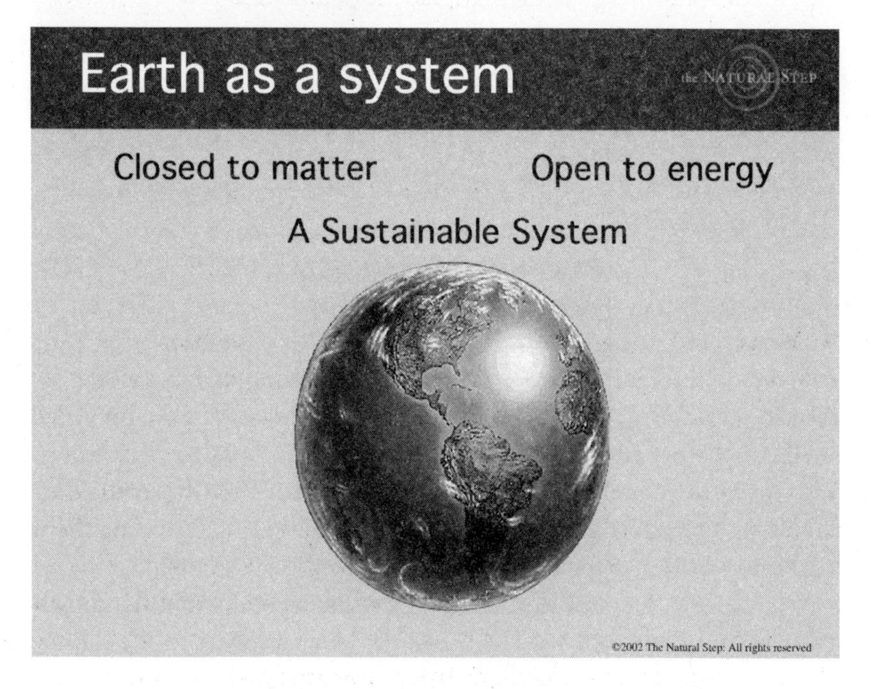

A series of events in the history of the cosmos has created the prerequisites for life on earth. The earth is a closed system (very little material comes into or goes out of this system). However, it receives light from the sun and sends out infrared radiation into space. In scientific terms it is the difference in thermodynamic potential between these two flows that has provided the physical condition that makes life possible.

The Natural Step refers to some fundamental science to help us see the bigger picture of how this system works.

Basic Science

1) Matter and energy cannot be created or destroyed.
   The 'Conservation Laws'. *The principle of Matter Conservation and the First Law of Thermodynamics.*
2) Matter and energy tend to disperse spontaneously. *The Second Law of Thermodynamics.*
3) Quality can be expressed as the concentration and structure of matter. *Economic value as well as material complexity.*
4) Green cells are essentially the only net producer of quality. *Photosynthesis.*

1. Matter and energy cannot be created or destroyed.

Our system, the earth, is a closed box as far as matter is concerned. Matter means any material or substance, solid, liquid or gas. Just about all the material stuff that was here when the planet began is still here. A few meteorites arrive occasionally and radiation leaks out, but these small comings and goings are relatively insignificant in terms of the make-up of the system. Matter changes form but it does not go away. At the same time this system is open to energy—light and heat from the sun. The system works; it is sustainable. And it goes on functioning through a wondrous cycle of creativity, the natural cycle.

2. Matter and energy tend to disperse spontaneously.

All the processes in this system keep on dispersing matter and energy, in a continuous cycle called entropy. Everything is in a constant state of breakdown towards its smallest molecular parts. Cars will become rust, carpets turn to dust. Energy eventually goes to heat, and matter loses structure. When that breakdown happens, the parts can be recycled, which is something nature does very well. Materials occasionally breakdown dramatically in volcanic eruptions and earthquakes, but otherwise they do so all the time through normal weathering. They are discharged and eventually tidied away again by natural cycles. Organic matter decays to be reabsorbed into the cycle, used as fuel and food by other organisms that will themselves break down in the ceaseless cycle of life. There is no waste in nature. Everything is in an ever-continuing process of breakdown and rebuilding.

3. Quality can be expressed as the concentration and structure of matter.

In the midst of this natural flux, life depends upon the way matter and energy can be held together in concentration, structure and purity to form a viable organism. We depend upon holding together matter and energy, for example, whenever we manufacture something for our purposes. What we are really doing then is creating a structure to hold things together. It is that structure that we consume, not the substance itself. For example when we burn petrol, what we are actually using up its quality and structure. We take the energy holding the fuel molecules together and their breakdown produces heat and material waste, in this case gases, which escape into the air. So we have not consumed the material, merely held together its structure for a while and transformed its energy content for our use.

4. Green cells are essentially the only net producer of quality.

We are only able to perform this transformation, this economic activity, because we have access to energy. The sun sends us much more than we can use. The problem is that raw sunlight is no good to us. We depend upon plants to convert that input from the sun into energy. Photosynthesis is the process that green cells use, which makes our air breathable. They are the ultimate producers of economic activity.

Where do we fit?

So we have identified some fundamental scientific knowledge that determines how things work in this system. Where do we fit? Well, human beings are part of this system, part of nature. We have needs for survival as individuals and as a species. To survive, we are entirely dependent upon nature's systems.

There is nothing particularly radical in this explanation of the system earth. It is fundamental science that has been known about for a comparatively long time. This is the starting point, the bread and butter, for environmentalists and ecologists. And in places where even the most basic education is widely available, most of the population will have been exposed to this knowledge at some time in their life. Yet do we live our lives as though we knew these fundamentals? Clearly not. It is as if something in the modern human mind rejects this knowledge.

I speak from personal experience. I freely admit that before this basic science was presented to me in context, I did not give it any real consideration as relevant to the issues around me. The market-driven world we inhabit does not tell us that stuff simply cannot be thrown away, because there is no 'away' to throw it! We have been conditioned to act and think as if we were set apart from and above the cycles of nature. The idea that we are actually 'renting' matter and energy, transforming it for our needs rather than creating and consuming, is not part of the way most people think.

The experience of The Natural Step has been that this knowledge can be communicated to individuals in a way that makes a difference. True learning, to the extent that is capable of changing behaviour, seems to require that the individual experiences insights at a very basic level. In any event, it is a precondition of The Natural Step Framework that we share a better understanding of the system within which we must all survive. That understanding cannot be approached in a meaningful way other than through science.

Chapter 4

Conditions for Success

How is it that we are so effective in threatening to overwhelm the cycles of nature? If this system has always been so resilient, why are we making such a difference to it now?

With a better awareness of the cycles of nature we can identify the fundamental ways in which our interventions are disrupting that system.

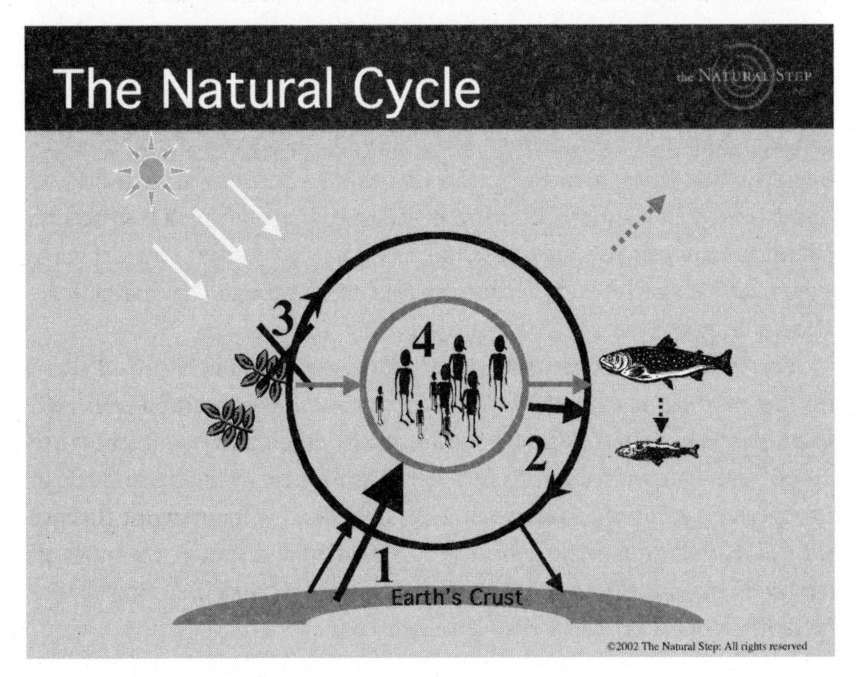

1. Breaking the cycle—digging it all up again

In the first place we have started to take so much material from out of the earth, that nature is just not able to handle it within its normal cycles (flow **1** in the diagram overleaf). Burning fossil fuels results in an accumulation of gases in the atmosphere at a rate that is changing the climate. Our mining activities are now releasing metals, including heavy metals, from the earth—originally themselves toxic gases which were sequestered into rock sediments and minerals over millions of years—into the biosphere at a dangerous rate. We know from the first science principle that this matter does not just disappear. Even if we can't see it any more when it is vented to the atmosphere or spread on the land, it is still there and accumulating.

The second science principle tells us that it will tend to disperse and spread, permeating an ecosphere that is used to relatively low levels of carbon dioxide and heavy metals. This systematic accumulation is bound to have consequences. It overloads nature's circular systems, and at some point our ability to produce structure and usefulness from nature's resources becomes more and more problematic.

2. Breaking the cycle: poisoning the system

The second big burden that we, people of the industrial age, are inflicting upon our home planet is to release into its atmosphere a whole load of new substances (flow **2** in the diagram). Around 750,000 synthetic chemicals are now on the market. Only a small fraction of these have ever been properly tested for their effects. We know that chemicals affect humans and often have disturbing propensities to alter hormones, immune systems and to damage the functioning of cells. Even if there were no concern about the direct risk to human health, we should still be concerned about this situation. Many chemicals in common use do not break down in nature, but they accumulate at different levels in the cycle. Others break down very slowly. Just like everything else, these substances do not disappear—they spread and disperse, turning up downstream in unexpected places. There is hardly a more controversial area of science than the debate about the possible links between chemicals and human ill health, including cancer. It is hard enough at the personal level to contemplate the implications of accumulated substances that have been called 'public poisons' (insect killers, industrial solvents, incineration residues). But the effects that are passed on to future generations are even more disturbing.

3. Breaking the cycle: destroying the engine

We are also in grave danger of destroying the part of the system that makes life possible at all. With every day that passes, we lose more of the productive surfaces of the earth (flow **3** of the diagram). We are covering over, burning, bulldozing, stripping and generally destroying forests, prairies, marshes and other kinds of terrain. The wrecking of ecosystems does more damage than most people realise. That is the way of systems. If you are careless enough to damage a part, the consequences will affect the whole. You can guarantee there will be consequences, even if you cannot predict exactly what they will be. The irony is that much of the time we do actually know what the consequences will be. We know that biodiversity is absolutely essential to so much of our economic capability. Shake a tree in the rain forest and thousands of examples of life will fall out of it, examples which we use to produce new medicines, and technologies. And of course the tree itself is part of the system, which turns the air we breathe into air that keeps us alive. Nature is like a savings bank, which has built up a capital fund. That fund is able to replenish itself. Provided we are sensible, nature will put enough resources back into the fund to make up for what we use. To continue the analogy: we can do fine when we live off the interest—the harvest. But we are now systematically destroying nature's capacity to keep on replenishing that fund. As the interest runs out, we are starting to live off the capital.

4. Breaking the cycle: impacts of unrestrained economic growth

If there is one paramount reason why we humans are making such a difference to the smooth operation of the planet's systems, it must surely lie in the pursuit of unrestrained economic growth (flow number **4** in the diagram). Human society has unleashed a combination of sophistication and expansion never seen before. The ever-greater urbanisation of our society has had major environmental impacts. Vast waste dumps have been building up. We see information systems spreading to all corners of the globe. Our ever-expanding roads are chronically congested. Energy grids are barely able to cope. We have waste materials too dangerous to deal with, weapons capable of mass destruction. We can deal with diseases that were once thought to be unbeatable, yet new viruses are posing even greater threats. All this activity continues to go on without pause, even as we see the warning lights. Our ingenuity seems to know no boundaries, yet we continually fail to connect the dots of the global picture.

The picture which changing natural cycles should reveal to us is one of a system in distress. The repeated environmental and social damage caused by unchecked economic growth is reported in the news almost daily. Social conflict and poverty is reported alongside failing agricultural systems, unfair trading practices, and forests turned to desert. Yet we have hardly begun to get to grips with the root causes of these problems. The need to see a clearer and more complete picture lies at the heart of what The Natural Step, using science as its guide, sets out to achieve.

The TNS System Conditions

The work of developing our framework for sustainable development has been pursued by looking at numerous impacts related to the unsustainable, linear processing of matter. The Natural Step aims to connect those impacts into a simple but irreducible set of requirements for societal design and human behaviour. When these requirements are not met, we are acting against what science tells us about the way the world works. So we can describe how a sustainable society would be different from our current unsustainable society. Thus:

THE FIRST THREE SYSTEM CONDITIONS

In the sustainable society, nature is not subject to systematically increasing . . .

— concentrations of substances from the earth's crust;
— concentrations of substances produced by society;
— degradation by physical means.

To avoid overloading natural cycles through these three mechanisms, society must satisfy more human needs per resource throughput than it does today. Otherwise, it will be impossible to avoid systematically increasing concentrations of natural and synthetic waste (System Conditions 1 and 2), and it will be ever more difficult to live off the 'interest' of natural capital (nature's production of resources and services) rather than depleting the stock of capital itself (System Condition 3).

So what do these System Conditions mean in practice?

System Condition 1

Particularly during the industrial age, human society has produced, and is still producing, a net input of substances taken from the earth's crust (the lithosphere) into the environment around us (the ecosphere), for example fossil fuels and metals. These flows are often large compared to natural flows. The biological risk from a metal (or other substance) is approximately inversely proportional to the natural occurrence of that metal (or other substance). Even a small absolute amount of a relatively scarce metal risks large relative increases of concentration of that element to which biological systems have been relatively unexposed.[3]

The ecosphere has a limited capacity for assimilating the intentional and unintentional flows of mined elements. Sedimentation and dilution operate slowly in nature relative to today's flows from human activity. For sustainability, the balance of these flows must be such that concentrations of substances from the lithosphere do not *systematically* increase in the whole ecosphere or in parts of it, such as the atmosphere or different ecosystems. These critical concentrations differ from one material to another and between different receiving environments. The complexity and delay mechanisms in the ecosphere may also make it difficult to foresee what concentrations will lead to unacceptable consequences. Therefore, the very least we must achieve is a halt to systematic increases of concentrations.

---

3. Azar, et al, *Socio-ecological indicators for a Sustainable Society*, Chalmers University, Gothenburg.

This System Condition does not say that we cannot use metals in society, or that we must stop mining. That would be nonsense. We have always depended upon these resources, and a sustainable society is unlikely to be possible without material taken out of the ground. It is the rate of extraction and, crucially, the management of such materials once extracted, that really matter.

System Condition 2

In almost all parts of the modern economy, society produces chemical compounds. They are the unseen camp followers of modern life. The problem is that these compounds are emitted, or leak out, into natural systems. System Condition 2 says that the flows of such molecules and nuclides to the ecosphere must not be so large that they can neither be degraded by, or integrated into natural cycles, or deposited into the lithosphere. The balance of flows must be such that concentrations of substances produced by society do not systematically increase in the whole ecosphere or in parts of it, such as the atmosphere or different ecosystems.

What concentrations can be accepted without jeopardising our health and economy in the long run depends on such properties as eco-toxicity and bio-accumulation. Eco-toxicity is the negative influence of a substance on ecosystems. Bio-accumulation is a measure of the extent to which a substance is taken up by living organisms and concentrated in the food chain. As with the previous System Condition, the critical concentrations differ by substance and recipient. It is often very difficult to foresee what concentration will lead to unacceptable consequences. Therefore, what must be achieved is at least a halt to systematic increases of concentrations.

This System Condition does not say that we cannot use chemicals in society or even such compounds that are foreign to nature. The use of some polymers, for example, can in many ways be very beneficial to sustainable development, and in helping to meet human needs. It does mean decreased turnover of substances that are increasing in ecosystems today (for example nitrogen oxides), and a phase-out of persistent substances foreign to nature (for example PCBs). It means better management and containment of substances.

System Condition 3

Human productivity has always involved physical impacts on natural systems. System Condition 3 implies that natural systems must not be systematically degraded by over-harvesting, mismanagement, displacement, or other forms of physical manipulation. There are quantity and quality issues here. The increasing deforestation of the earth's surface, or more and more coverage of that surface by human activities, all lead to a systematic reduction of green areas that regulate our climate. Equally, manipulation and interference with ecosystems can seriously damage nature's capacity to provide the resources we need, and to deal with our waste. In practical terms in today's situation, this means changes in our practices within such areas as agriculture, forestry, fishing, urban planning, infrastructure, construction and manufacturing.

This System Condition does *not* say that we cannot use nature's resources. Without any manipulation of nature, only a fraction of today's human population could be sustained. So it may be necessary to build a road on fertile land; that is not necessarily a sustainability problem. It is more and more roads on fertile land that is heading towards un-sustainability. Of course we need to harvest crops and all manner of natural materials. It is the consequence of systematic depletion of ecosystems, increasing monocultures on land suited to diversity, introducing organisms to foreign habitats, which should cause us concern.

Today society is violating these System Conditions. Waste is steadily accumulating and resources are diminishing. The decisions we make need much more care and consideration. Sustainability risk assessments, and paying attention to the systems impact of decisions, need to become commonplace.

Conforming to the three System Conditions for ecological sustainability is a prerequisite for being able to fulfil human needs in the long run. If we are more efficient technically, organisationally, and socially, we can provide more services. There is then the possibility of more humans, now and in the future, being able to meet their full needs, and enjoying a decent way of life.

In Chapter 6 we will dig deeper into System Condition 4, the social dimension of sustainability: **In a sustainable society people are not subject to conditions that systematically undermine their capacity to meet their needs.**

"The Natural Step should stand as a beacon to us to understand what is really going on in the world and to begin to take steps to correct it. There is no human effort going on now which is in closer touch with reality, which is more necessary for us or which affords better hope for our common future."—Dr. Peter Raven, Director, Missouri Botanical Garden; President, American Association for the Advancement of Science; and past Secretary, National Academy of Sciences, USA

Chapter 5

The Natural Step Framework

Strategic processes for organisations to make progress

The System Conditions, and the way in which they are derived from science, are perhaps the most widely known feature of the TNS Framework. Over the years TNS has also developed a number of other methods of addressing sustainability.

In this chapter the most prominent of these methods are briefly outlined. They all owe much to the systems approach. You will not find any DIY sustainability kits here. Every organisation, issue and process needs to be dealt with according to its own particular circumstances. And, if action is to take place towards sustainability, it also needs to follow some key principles—it needs to be strategic.

> *"The whole world has dreamt about a solid definition of sustainability that would allow systematic step-by-step planning. When that definition arrived, delivered by The Natural Step, it was remarkable to see how simple it was. Why hadn't anybody thought about it before?"*—Paul Hawken, author, *The Ecology of Commerce*; co-author, *Natural Capitalism*

The Funnel Metaphor

The systems view reveals an important reality of today's unsustainable society. The problem of un-sustainability is not only that we have emitted a lot of pollutants causing some impacts. The problem is that industrial society is designed so that the concentration of pollutants is bound to increase globally. For example, it follows from the laws of nature that as long as energy systems are organised as they are, atmospheric concentrations of greenhouse gases will continue to increase, contributing to climate change. At the same time natural systems are systematically deteriorating due to destruction by physical means, such as over-harvesting and growth of industrial and urban infrastructure. In short, waste is steadily accumulating whilst resources are steadily declining. Therefore, the resource-potential for society and the economy is systematically decreasing. At the same time, the earth's human population is increasing and the gaps between the haves and the have-nots are growing. Therefore, unsustainable development can be visualised as society entering deeper and deeper into a funnel, in which the space for deciding on options is becoming narrower and narrower per capita.

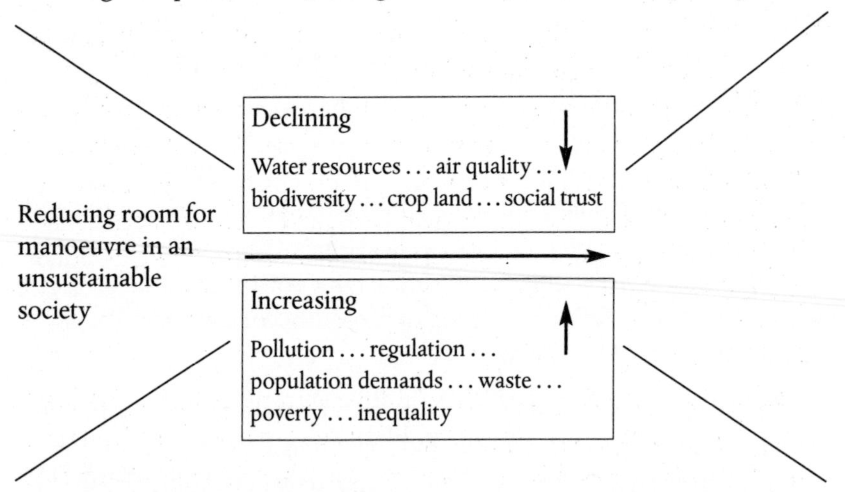

Hitting the walls of the funnel for any organisation may appear as:

- Decline in resource availability, biodiversity and social stability
- Increases in costs of pollution, waste management, taxes
- Ever stricter legislation and higher insurance premiums
- Lost investments
- Loss of good reputation

You cannot change an organisation's dependence on energy systems, infrastructure, technologies and management routines overnight. What we do today influences our chances tomorrow. The funnel is a metaphor that is applied to sensitise businesses, communities and society at large to the bigger picture. It is a part of our approach which works well at all levels. You can ask an organisation to think about the big global issues and also to draw their own funnel. What are the growing constraints on their own ability to make progress? Identifying the constraints is the first important step to dealing with them. Once again this method helps organisations and individuals to think in a smarter, more systematic way.

Development that is sustainable takes into account all the constraints and limits. It avoids hitting the funnel walls and aims to create new options for the future. We refer to the opening-up of the funnel walls as reaching sustainability, when our demands upon nature's resources are met without destroying the system's capacity to go on meeting those needs.

Backcasting

Creating visions to guide sensible planning has become part of the TNS Framework and we refer to it as 'backcasting'. It is really a very simple concept but has great success in unleashing creativity amongst those who use it. It is of course so named to place it in contrast with the more familiar exercise of forecasting!

Backcasting means: *placing ourselves in the future and imagining that we have achieved success. Then we look back and ask the question: "How did we achieve this?"*

The crucial starting point however, and the reason this lends itself so well to the TNS methodology, is that we must first of all understand the system in which we operate and the limits, or conditions, for success.

Backcasting sometimes works best, for an organisation, when people can be encouraged to really let go of all the problems and constraints they are normally working with. Groups do it best when encouraged to be imaginative together: what would this company, this local authority, this community, look like in a sustainable world? What would we be doing, how would we be dealing with materials, energy, transport, location, and relationships? The detail comes surprisingly easily when we think about being in a state of success and being part of a sustainable society. You can use any timeline so long as it is far enough away to stop people thinking about today—25 or 50 years work equally well.

When that vision is constructed then it can start to energise the crucial step of action planning. If we were successful, what were the obstacles that we managed to overcome? What were the video, the press release, the breakthrough investments, and the milestones that told the history of our organisation's journey into sustainability? What were the crucial elements in the wider society that had to change in order for us to make it through the funnel? It is particularly important to identify external factors, no matter how intractable to change they may seem. It is often the case that organisations do not rate their capacity very highly to get involved in promoting change in wider society. Often they will say that new laws or regulations, more cooperation or better informed customers, are necessary for them to make any progress, as if that is an excuse for doing nothing. When backcasting is done well it gets people to see that there is always more they can do, and that their contribution can make a real difference.

Spotting the obstacles to sustainable development and imagining how we overcame them is a first vital step in constructing a sustainability strategy and action plan. The method was not invented by TNS but adapted and attached to our principles for success. Backcasting without such principles is not really much use to us in our quest for a sustainable future.

The problem of traditional forecasting, put simply, is that when we are doing it we tend repeatedly to take today's problems and preoccupations with us into the future.[4]

The ABCD Process: A Step-by-Step Strategy

In Chapter Four we set out the Conditions for success, defining critical features of a sustainable human society. Each individual organisation must draw its own conclusions from these basic principles as regards problems, solutions, goals and sub-goals. The four-step 'A,B,C,D'

---

4. Holmberg, J., And Robèrt, Karl-Henrik, 2000, 'Backcasting from Non-Overlapping Sustainability Principles: a framework for strategic planning', *International Journal of Sustainable Development And World Ecology*, volume 7, pps. 1-18.

process below provides a systematic way of guiding them. At TNS we have found this systematic approach really helpful in our work. It sets out the basic milestones that we would urge any organisation to follow in seeking to become more sustainable. It is not a rigid formula. It does not have to be followed in this order. However it is our experience that each of these activities is going to be needed at some point, no matter how large, small or sophisticated the organisation may be.

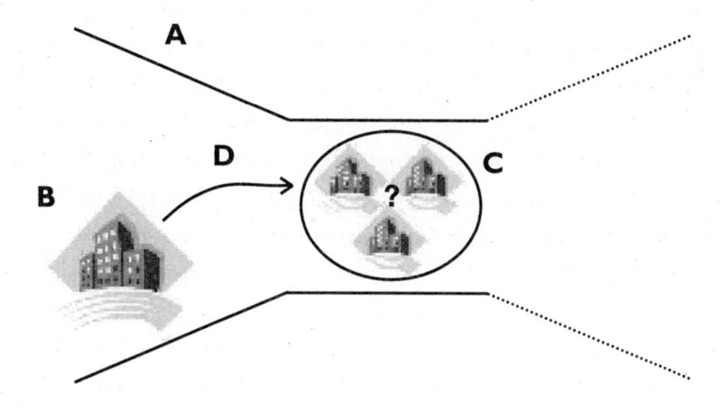

STEP A: Awareness: The starting point has to be a much better understanding of what is going on around us. We need to be aware of the bigger system within which our strategies operate. We will never be able to plan adequately unless we take account of the full consequences of our actions, and the consequences can only be assessed if we see the wider system more clearly. Long gone are the days when any organisation could act as if it were only responsible for those things that it directly controlled. The world is much more complex than that, and we need decision-makers and policy-makers to take responsibility for all the impacts of their actions, direct and indirect. The TNS Framework—including the System Conditions—is designed to help in understanding the wider scene. It sets out very clearly the things we must stop doing if we are to move towards a sustainable society, and explains why. The key learning issue here is that we cannot expect anyone to sign up to new and better ways of behaviour if we do not give them the reasons why. We have to enroll people into this sustainable development effort with their hearts and minds. Everyone does things better if they know why they are doing it, rather than just being told that they must. It is only through building a shared awareness of the fundamental importance of sustainable development that we will begin to make progress.

STEP B: Baseline: An assessment of 'today' is conducted by listing all current flows and practices that are problematic from a sustainability perspective, as well as considering all the assets that are in place to deal with the problems. Once organisations are aware of the underlying principles for a sustainable society (the System Conditions) they can evaluate their own performance. Without that understanding, the evaluation has no anchor. An assessment of where we are today, with that better understanding, will reveal many practices, resource flows and impacts that would not otherwise have been considered in the sustainability context. The System Conditions can be used to evaluate every activity in all organisations.

STEP C: Visioning: What would this organisation, this product or this policy, look like in a sustainable society? That key question should underpin every strategy. The process of backcasting helps to produce the answers. Solutions and visions for 'tomorrow' (i.e. the opening of the funnel) are created and listed by applying the constraints of the System Conditions, to scrutinise suggested solutions and trigger creativity.

STEP D: Setting and Managing Priorities: Priorities for action, new investment and new strategies, are selected from the C-list. However far ahead the vision may be, we have to make a start. In an ideal situation we would pick smart early moves. Actions and changes that give identifiable and beneficial results are important early on. Whatever decisions are made can now be judged with better understanding of the bigger system, and with a vision of success to guide them towards sustainability.

Each measure should be tested against the following questions:

- *Does this measure proceed in the right direction with respect to all System Conditions?* Sometimes a measure represents a trade-off that proceeds in the right direction with respect to one of the system conditions while working against others. Asking this question helps illuminate the full picture, and lead to complementary measures that may be needed to take all System Conditions into account.

- *Does this measure provide a stepping-stone for future improvements?* It is important that investments, particularly when they are large and tie resources for relatively long time periods, can be further elaborated or completed in line with the System Conditions in order to avoid dead ends. An example would be investing heavily in a technology that may result in a few less impacts in nature now, but will not be capable of later adapting to contribute to complete compliance with the System Conditions.

- *Is this measure likely to produce a sufficient return on investment to further catalyze the process?* It is important that the process does not end due to lack of resources or bad investments along the way.

Measures that answer 'yes' to all three questions provide the strategic element of the methodology.

The Five Level model for planning in complex systems

Systems thinking does not seem to come easily to groups, particularly when faced with difficult or complex issues. As individuals we may all be good systems thinkers, but in groups this capacity can get lost. Our attention focuses on our individual concerns, and seeing the bigger picture is a greater effort. This at least has been our experience in TNS, and we have needed to work harder to improve the quality of systems thinking amongst people we work with. One of the ways we do this is to use the Five Level model set out below. An immediate benefit comes from everyone being able to clarify the level they are talking about. Too often, when we are dealing with complicated topics, we find that misunderstanding comes from people talking about different things without realising it. The model we have adapted from various sources and from our own internal experience helps to overcome that problem. In any dialogue about sustainable development we can check whether we are talking about the same levels.

LEVEL ONE : The System
Here we describe the system itself. We look for the big and important flows and connections within that system that are essentially what makes it work—the system dynamics. Science helps us understand those dynamics.

LEVEL TWO: Conditions for success in the System
Knowing the fundamental requirements of how the system works means we can identify the conditions for success—TNS System Conditions. We can build a vision of how we would need to act in that system to be sustainable.

LEVEL THREE: Strategic Guidelines
We can set out some clear guidelines, common for all the decisions we might face, which if adopted will lead us in the right direction.

LEVEL FOUR: Actions
The things we can actually do to make progress towards sustainability; actions that result in sustainable solutions, actions that build our capacity and actions that help us to learn from and to evaluate our progress.

LEVEL FIVE: Tools
Whatever the context or the purpose there are almost always tools that can help. We can learn form the actions of others and take advantage of the many techniques, models, procedures, and measuring tools that are available.

Together with the System Conditions, the funnel, backcasting, ABCD process and the Five Level model are all part of the TNS Framework. They have evolved from our work with organisations in trying to disseminate the message of sustainability and to promote practical action. The next chapter examines System Condition 4 in more detail. It is an area of our work that has received more attention over recent years.

Chapter 6

Social Sustainability

TNS System Condition 4: In a sustainable society people are not subject to conditions that systematically undermine their capacity to meet their needs.

How do we understand better the connection between human society and nature's systems?

From a sustainability perspective, there is a direct correlation between the social and ecological dimensions:

- On the one hand, social sustainability's dependence on wider ecological sustainability is becoming more evident. As we continue to undermine nature's capacity to provide humans with services (such as clean water and air) and resources (such as food and raw materials), both individuals and the social relations between them will be subjected to growing amounts of pressure. Conflict will grow and public health, personal safety, and other negative social factors will increase in the face of ecological threats and decreased access to nature's services and resources.

- On the other hand, overall ecological sustainability has become dependent on social sustainability. If a growing number of people are living within a social system that systematically constrains their capacity to meet their needs, then participation and investment in that system will break down. The end result of such socially unsustainable development is rising violence, alienation, and anger. People will place no trust at all in nature once social trust collapses and various modes of barbarism develop. Conflict, poverty and other forms of social stress will result in more environmental degradation.

These dynamics—ecological threats leading to social unrest resulting in greater ecological threats—are the basis of deeply unsustainable patterns of behaviour in which humans are currently enmeshed. They illustrate the indivisibility of the system.

Regardless of where an individual or organisation places the primary focus of interest in the sustainability debate—at the social or wider ecological scale—the fact remains that the social questions are the key leverage points. Human behaviour and the resulting societal dynamics are the basis of the current social and ecological problems. In this chapter the Five Level model for planning in complex systems is applied to social sustainability. This is the result of considerable work inside TNS to arrive at guidelines for working with the fourth sustainability principle. This work has been undertaken at the same time as we have seen a growing interest in Corporate Social Responsibility and more questions being asked about building sustainable communities.

In the business sector there has been a growing trend towards Corporate Social Responsibility (CSR) over the past decade. Companies are increasingly aware of the demands upon them from the social welfare agenda. They face forces potentially in conflict with business interests. Consumerism, health concerns, environmental concerns and attacks from the anti-capitalist/anti-globalisation movements are now serious issues in boardrooms around the world.[5] We believe that undoubted benefits of CSR in business will only be realised in the longer term when it is firmly rooted in sustainable development policies.

It is not only in the business sector that we see a need for a fresh and better-informed approach to social issues. Decisions that directly affect the lives of individuals are made in a complex array of gatherings and institutions. Multilateral agreements on trade, the movement of people, defence, food and health have implications for us all. They impact daily upon our capacity to lead more fulfilling and sustainable lives. Decisions and policies that affect our social fabric and our relationships

5. Echo Research, *Giving Back: integrated research into CSR in global markets, 2000/2001*, www.echoResearch.com.
Caulkin, Simon, 'Can Business Save the World', *The Observer*, London, 5 Oct 2003.

need to be considered in the context of sustainability. The same is true at the local level. In municipalities, community groups, and all the organisations which have a role in the way society works, their job needs to be done with much more regard to the imperatives of sustainability.

> *"The real voyage of discovery consists not in seeking new landscapes but in having new eyes."*—Marcel Proust

In TNS we believe science and systems thinking, and the limits they reveal, can assist us with the social dimension. Are there dynamics, or mechanisms, in nature's systems, which can guide us in the tricky zone of our own social system?

Interdependence, self-organisation and diversity are features that all living systems have in common, from the most complex to the simplest.[6] We have set out the importance of these mechanisms as guidance to the way systems work.

> *"A system is an interconnected set of elements that is coherently organised around some purpose."*—Donella Meadows

We need to be mindful of these intrinsic mechanisms (design principles) when thinking about the way we organise our society and our impacts upon social groups. We cannot 'adopt' these concepts, anymore than we can adopt thermodynamics; they just are. It will never be helpful to think about people as above and beyond evolution and nature itself. And even if our religious beliefs were to tell us that evolution has not occurred in line with the scientific myth, but is entirely an act of God, this does not change the constituents of the system. If it was not Big

---

6. Wilson, E.O, *Consilience,* 1998; Anders and Bromm, *Profit without Measure,* 2000; Capra, Fritjof, *The Hidden Connection: a science for sustainable living,* 2002; Williams, George C., *Plan and Purpose in Nature,* 1996; Zohar, Danah, *Rewiring the Corporate Brain: using new science to rethink how we structure and lead organisations,* 1997; Meadows, Donella, *Places to Intervene in a System,* Whole Earth, 1997.

Bang, but God who created the laws of thermodynamics, this does not affect the rest of the reasoning in this chapter, or any other chapter. We are still a part of that system and we are governed by the same system dynamics.

We know of no philosophical or religious belief system that would tell us not to take responsibility for nature or ourselves—the system on which we rely. On the contrary, we have growing experience from organisations such as the UN, and its work with the Universal Declaration of Human Rights, that it is possible for people from different cultures to co-create generic guidelines for human behaviour on the international scene. So, how do we uphold a dignified social covenant on the global level, and how can we become more strategic about it?

"A genuinely fundamental and hopeful improvement in 'systems' cannot happen without a significant shift in human consciousness, and … it cannot be accomplished through a simple organisational trick. This is something no revolutionary or reformer can bring about; it can be only the natural expression of a more general state of mind in which man can see beyond the tip of his own nose and prove capable of taking on responsibility even for the things that don't immediately concern him, and relinquish some thing of his private interest in favour of the interest of the community, the general interest. Without such a mentality, even the most carefully considered project aimed at altering systems is for naught."—Vaclav Havel

Level 1: The Social System

Nature (i.e. the biosphere) is a system in which all species—including humans—are interrelated. As a result of our interdependence within this system, every action has consequences and secondary effects in other parts of the system effecting other human beings, social groups or species. Effects 'ripple out' from every action.

Un-sustainability within this system is about the tension between conditions for ecological sustainability on the one hand (System Conditions 1-3), and conditions for social sustainability on the other (System Condition 4). Un-sustainability is not related to some external threat or catastrophe. It is human behaviour that causes the problems. We must learn to develop and prosper within the social and ecological limits of the system. We will not attain a sustainable relationship with nature unless we attain sustainable relationships with each other. We cannot take care of nature until we take care of each other. Sustainable development is about taking care of human needs without systematic damage to natural systems. By human needs we mean inborn requirements that need to be satisfied in order for people to remain *healthy*—physically, mentally and socially.

There have been many excellent attempts made at understanding human needs, from the most basic to the more complex. A well-known example is the definition of nine distinct human needs by the Chilean economist Manfred Max-Neef,[7] and his categories are listed below.

THE MAX-NEEF CATEGORIES OF NEED

Subsistence	Protection	Participation
Idleness	Creativity	Affection
Understanding	Identity	Freedom

If you think about these nine categories carefully you will see that they are indeed essential to us all, and intrinsically related. We can approach this by imagining—'what would happen if I were to be completely deprived from any of those'? For example every individual needs affection. Without the care, and sometimes love, of others we are diminished as human beings. There are consequences for all of us whenever we are systematically deprived of the opportunity to meet these individual needs. And on a global scale, those consequences are stark.

Needs are fundamental to the quality of life for all of us. Max-Neef states that if one of the human needs is in systematic short supply, this leads to 'poverty', regardless of how much money the individual in

7. Max-Neef, Manfred, Development and human needs, in Ekins, Paul, and Max-Neef, Manfred, eds., *Real-life Economics: understanding wealth creation*, Routledge, London, 1992

question happens to have in the bank. If nothing is done to remedy the shortfall, the individual runs the risk of developing various 'deprivation symptoms/diseases'. These types of deprivation symptoms/diseases are widespread and not only in 'developing countries' but also the industrialised countries. 'Economic growth' does not function as a defence or guarantee against these effects. Examples may include various kinds of stress-induced diseases caused by a lack of *idleness,* alienation caused by a lack of *participation,* and insecurity when we do not have *protection* from violence or terrorism, which may themselves result from someone else's needs not being met. Overall global economic growth that is, in its current form, running up against ecological limits, has not been able to meet the human physical needs either—over one billion people on earth are starving from a medical point of view.

Society: the system and its goals

There is an important distinction between the capacity to meet our individual needs and the *opportunity.* We are born with the capacity, but we are utterly dependent upon others to provide the opportunity for us to exercise that capacity. We form groups and communities to enable us to meet our needs. That is the origin and purpose of society and all its component parts sometimes referred to as Social Capital. This stock of capital includes social networks, laws, schools, governments, clubs, agreements, religions, business organisations and health-care systems and all the social arrangements that make things work. The ultimate goal of society is to meet individual human needs, with all of the complex balancing and judgement that such a responsibility entails. The effectiveness of its social capital in achieving that goal is how any society should be judged and valued.

Today, however, the progress of most national societies has become synonymous with growth of Gross National Product. This is a measure of a nation's total consumption and investments. Yet there are no universal linkages between the size of those resources flows and the life quality of the respective populations. Therefore, we need to develop new measures that take the supportive system ('natural capital') and the human system ('human and social capital') into account. It is worth remembering that the term 'economy' refers to making the best use of scarce resources. Economic efficiency cannot be evaluated unless the ultimate utility (meeting human needs) is considered. It is no longer acceptable to be concerned only with quantitative resource flows ('manufactured and financial capital'). We also need to attribute value to qualitative resource flows. Healthy and resilient ecosystems and societies are becoming our new and dominating scarcities. In order to raise long-term welfare and quality of life all over the world, most of the world's countries will need to develop approaches (including organisations, institutions, technologies and new methods of monitoring) that are different from those that are dominant today. The new paradigm—a global 'taking care of the planet by taking care of each other' culture—must be implemented in partnership across all national and community boundaries.

We already see increasing numbers of endeavours around social sustainability from all kinds of communities around the world, municipalities, universities, businesses and individuals, where these challenges are being taken seriously. So the alternatives already exist and are growing in numbers and scale. Our main problem is that we are in a hurry and these examples are simply not mainstream. Unsustainability inherently means that we are losing not only ecological resources such as fresh water reserves and biodiversity, but also social resources such as cultural diversity, and societies with 'stories' of meaning that tie people together and create trust and lust for living together.

Social organisation

We build social relationships, and therefore social capital, in order to meet our needs, both individual and collective. How could we gain insight into some constitutional aspects of social capital to guide us on the organisation level? To help us do that we can turn to systems in nature. Nature is the basis, or foundation, of all systems. Everything works, everything connects and everything has its place and time in nature's system. It is not difficult to see that there is some internal design in assembling ecosystems. There are three characteristics that are widely referred to in science as determining the efficiency, or fitness for purpose, of all systems in nature, including social systems. Interdependence, self-organisation and diversity are the overriding features that all living systems have in common, from the most complex to the most simple. These concepts do not overlap, they work together, and cannot be substituted one for another. They provide the context and reason behind many other interactions. For example the adherence to individual human rights, to take a very human example, arises from both the self-organising capacity of every diverse individual to achieve certain standards of living, and their mutual dependence on the wider social system to enable them to fulfil that capacity.

If we are to achieve social sustainability, we must identify nature's limits both in physical resource flows (*which determine quantity*) and in the pre-conditions for living systems (*which determine quality*). The following three sections describe these interrelated characteristics in more depth.

Self-organisation

Since all components of life (cells, organs, bodies, species in the ecosystems) are dependent on each other, evolution has brought about an individual capacity of those components to self-organise within that context. Ecological systems rely upon the self-organising capacity of their individual parts. Those parts have the physical, chemical and biological capacity to meet their needs. We observe this capacity in ant colonies, flocks of birds and the cells in our own bodies. Each individual in these natural systems knows what to do. The functioning of the system is dependent upon the respective parts having the opportunity to exercise that capacity. Likewise, people have an in-built capacity to organise themselves in ways that increase their prospects of satisfying their needs. Each person has the capacity for participation, creativity, to give and receive affection and understanding, and thereby establish their identity within the group. To make this possible, we organise ourselves in complex social structures. The characteristic of meeting our individual needs through social structures is one we share with many other species in nature.

For human beings, empathy has a basic constitutional importance when it comes to our capacity for self-organisation within the human society. Empathy is a concept free of any moral implications. How and for what purpose we use that capacity, or whether we do so at all, is another issue. Empathy simply describes a particular in-built human capacity for detecting even the subtlest of signals from other people and entering into their situation. It helps us build relationships and social structures, crucial for us to be able to satisfy our needs—not just social needs, but all needs. We have an obligation, as responsible citizens, to use that capacity in ways that do not hinder social cohesion. Conversely, society has an obligation towards us. Social structures which deny individuals the opportunity to use their in-built capacity can never be sustainable. Decisions we make at the organisational or political level can result in people being denied access to resources, to education and to quality social capital. This prevents them from exercising their self-organising potential to lead fulfilling lives. Their social deprivation becomes our terrorism. Their ecological deprivation becomes a problem for us all. Therefore, the preservation of individual human rights and responsibilities should be a paramount consideration.

Diversity

We also learn from nature that diversity is a key component of any living system. The changes which take place in a forest eco-system involve many stages. Diverse species are seeking to find their place, using the resources available to them in that system, and mostly do so with an amazing efficiency. This makes up the ecosystem's 'biological diversity' of which they are part. Diversity and its accompanying efficiency are critical to the dynamics of that system. Monocultures do not survive. Resilience of the system comes from its ability to accommodate and promote diversity. The self-organising capacity of our own cells produces many different outcomes both within our bodies, from elbows to eyelashes, and between us, from Africans to Eskimos. At a human level a basic characteristic of strong, social structures is their ability to utilise *diversity*—the reality that every individual is unique. Societal and cultural diversity lie at the very foundation of our species' brilliant ability to adapt and survive in any given situation.

Acting together and employing empathy in multiple ways, we are able to see more and know more and perform better than any individual could by himself or herself. But there is more to diversity than just working together and utilising our comparative advantages. It means unlocking the creativity in all of us by allowing different approaches to flourish. It means that by tolerating many cultures and traditions we are all enriched. All people respond to their circumstances in their own way, and that will be adapted to the climate, geography and social group in which they find themselves.

Interdependence

We also need to take heed of that feature in ecosystems which makes the optimum value out of self-organisation and diversity for the benefit of every system—interdependence. Just as the climate of the earth constantly responds to feedback signals and achieves an incredibly fine balance (dynamic equilibrium) of the overall system, so every system is affected by the actions and responses of its component parts. The renegade cell (e.g. cancerous cell) in an organism produces a response from the body. A predator will maintain its territory in relation to the territory of others. So each component of life (cells, organs, bodies, species in the ecosystems) not only strives to survive but to do so within the context of the collective needs. An individual is not only made up by cells, but also provides the joint habitat for each one of them. Darwin didn't say: survival of the fittest, but survival of the *fitting*. Over-exploitation of resources by any species will lead to the demise of the ecosystem, and therefore ultimately to its own extinction. The fact that we all find ourselves in the same big system means that we must find ways of working together, just as cells are equipped to work for themselves *as well as* for the whole body. We are interdependent, upon each other as well upon ecological systems. Whilst there is always a portion of the human race that is denied the opportunity to meet its needs, we will never be free of social and ecological devastation somewhere in the world. All successful natural systems balance the self-organising capacity of diverse individuals and groups with the overriding principle of interdependence.

The difference between the interdependence of species and ecosystems, and the interdependence between people and their environment, lies in choice. A human being can choose whether or not to comply with the rules of the system, a cell cannot. The choices that are available to us involve the exercise of power, by individuals but also by organisations. A systematic imbalance of power within the system, power that is exercised to the detriment of others and is blind to interdependence, is unsustainable.

Conclusions from natural systems

So the exercise of self-organising power, which then denies others the opportunity to use their own capacity to meet their needs (e.g. slavery), can never be part of the sustainable society. Power accrues to individuals and to organisations in a variety of ways. We accede power to institutions to act for us, we buy goods and services, which give the purveyors of those goods and services a degree of power and influence. Individuals who are making better use of their self-organising potential will become 'leaders' or have greater access to resources. Circumstances and history will often dictate where power lies. Whatever the source of such influence and power, it lies within a system and so carries with it a responsibility to recognise the implications of its use and abuse.

Equally, diversity per se can impair the functioning of systems when it produces cultures which deny individuals the basic opportunity to lead fulfilling lives. Such traditions, which constrain individual human rights, will create obstacles to a functioning social system.

Self-organisation and diversity therefore need to be acknowledged for what they are—intrinsic components without which society will never survive. They should be celebrated as giving us the capacity to lead creative, colourful and interesting lives. But for society to achieve sustainability, these components also need common acceptance of the interdependent nature of each of us on the planet. The translation of those mutual dependencies into effective relationships is the purpose of society. For that, we all carry responsibility. In short, a well-functioning social system depends on 'rights with responsibilities'.

Since we are *interdependent*, we need to be able to exercise our inborn ability to self-organise in that context, and—to that end—utilise the inherent diversity of life, including the diversity of the human society. This is crucial to the satisfaction of our needs—i.e. keeping ourselves physically, mentally and socially healthy. We should single out i) interdependence, ii) our self-organising capacity; iii) diversity; *and the interplay between these qualities,* as founding concepts in the creation of robust and sustainable human societies.

Level 2: Conditions for success of the social system

The fourth basic principle defines a sustainable society as follows:

In a sustainable society people are not subject to conditions that systematically undermine their capacity to meet their needs.

We have identified that all human beings have intrinsic needs and that the goal of the social system is to provide the opportunity for all to meet those needs, as a precondition to a dignified way of life for everyone. In the previous section, we arrived at the conclusion that interdependence, self-organisation and diversity are intrinsic components of robust and sustainable systems.

This means that any society will be ineffective in meeting human needs if it fails to secure political, cultural and economic systems that:

- Protect and promote human rights, freedom of expression, equal opportunities, freedom of movement and equal protection under the law (**Self Organisation**);
- Protect and promote freedom of thought, cultural and religious diversity and the freedom to be different (**Diversity**), and
- Protect and reinforce the kind of cohesion, reciprocity, trust and freedom of association, access to resources and education, on which all societies depend (**Interdependence**).

Since our current social course is unsustainable, we must identify the mechanisms through which people's ability to organise themselves into effective social structures are being systematically undermined *in spite of* the existence of our generally endorsed international codes of conduct.

So how can we identify systematic, generic and universal mechanisms that jeopardise the constitutional requirements of a sustainable human society? We believe that the term 'power' is essential to this. As citizens we should be most concerned with the behaviour of those who are in a position to undermine our opportunities to meet our needs. Governments, corporations, community organisations, of all shapes and sizes, exercise their authority on behalf of us all. Whether that authority comes through democracy, the market place, willing or unwilling agreement, organisations exercise their power as part of the structure of society. Corporate and political rights bring responsibility to those organisations, just as much as they do for the individual citizen. Failure to move towards social sustainability arises when organisations misuse power and influence. Such abuse of power ignores the realities of self-organisation, diversity and interdependence. The result is a systematic undermining of the capacity of others to meet their needs.

Progress to a sustainable society means examining the ways in which power is exercised in the network of human communities. For example the abuse of political power is a major obstacle to sustainability (political is here taken in its widest possible sense, derived from 'policy' such as 'organisational policy'). It may be government-sanctioned measures that run in the face of our conditions for success cited above. Other examples are closer to the individual organisational level, e.g. discrimination; exploitative supply contracts and exploiting local communities. An overriding mechanism for social un-sustainability is restriction on access to information for all. Failure to promote educational opportunities is a major contributor to social deprivation.

In modern culture the abuse of economic power is also often held to be an impediment to sustainability. Global corporations are accused of the kind of marketing that smothers cultural diversity, and feeds our materialistic desires and addictions. Abuse of economic power is possible in many different ways. Financial agencies may be restricting access to capital or to income-generating opportunities. Organisations with big buying power can dominate a supply chain. Subsidies in trade agreements may enable developed world enterprises to restrict growth and innovation in the developing world. Businesses with vested interest in their products can use their economic power to impede investment in more sustainable technology. Any organisation that abuses its economic position can result in others being prevented from meeting their needs.

Degrading the environment can be seen as another example of organisations, and the people within them, abusing their power. The pollution of communities, failure to provide workers, families and communities with safety protection or health care, are the result of organisations ignoring the responsibilities which go along with their powerful positions. This is not confined to big business, but also found amongst Government agencies and at the local as well as global scale. The ways in which organisations can produce stress for individuals and for their environments are numerous.

The conditions for success, in terms of achieving a sustainable society, must include an end to the abuse of power in these and other ways.

Level 3: Guidelines for successful strategies for social sustainability

On this level we look for some fundamental guidelines for social sustainability. They should underpin policies and strategies to avoid harmful social impacts. They are set in the context of achieving success within the system of human society.

For the third level of planning—the strategic level—we have found a comprehensive method for working with social and cultural impacts within organisations, with the aim of 'humanising' their decisions. 'The Golden Rule' is so named because it lies at the foundation of many ethical principles, and is even present in the world's religions. In the Jewish and Buddhist traditions, The Golden Rule is described in the negating form, i.e. 'You should not do to others what you do not wish them to do to you.' In the long term, avoiding doing to others and to society at large what we do not want others to do to us will be in our own self-interest. The Golden Rule is built on the idea of empathy—our inherent ability to enter into other people's situations. If we have this ability, why is it not automatically put in place? It is possible that it only works autonomously when people are actually meeting each other, and in general in communities that are relatively small such as in families and perhaps villages. Putting our capacity for empathy into effective use is more challenging in a global society. Our decisions impact upon people to whom we cannot put a face, or we may not know by name. This is even more reason to ask what The Golden Rule means for an organisation.

We must at least get rid of destructive influences on other people that cause us, and them, to be un-sustainable. Organisations need to be specific about eliminating possible political, economic, environmental and other forms of abuse from the consequences of their actions. Consideration of any policy, product, marketing or investment, should always identify in advance the people who are going to be affected, taking the widest possible systems view. And those responsible should ask themselves: 'Would we like to be subjected to the conditions we create?' Power exercised responsibly should result in policy and strategy, products and services, without harmful consequences for people or the environment, now or in the future. No development is capable of being classified as sustainable development unless it can pass that test.

As well as examining the impacts of individual strategies, progress to sustainability must also involve improvements in the way organisations arrive at those strategies. Removing the abuse of economic, environmental and political power from society and the relationships it contains, means giving much more attention to process. The dynamics of the system demand that. What may seem to be the most benign public policy, and the most socially acceptable business services, could still be unsustainable if they are devised or administered in an unsustainable manner. Those concerned with governance and decision-making structures need to ask questions about process:

- **Participation:** 'Is this decision based on enough participation from, and dialogue with, all affected parties?' (Would this degree of participation, if I were subjected to it, be acceptable to me?)
- **Transparency:** 'Is this decision planned in a way that has enabled people to gain access to information and monitor the process throughout?' (Would this degree of transparency be acceptable to me?)
- **Responsibility and accountability:** 'Has the responsibility for the decision been clearly communicated between all the people taking part in the planning process, including those who are affected even indirectly by it?' (Would this degree of clarity in terms of responsibility and accountability be acceptable to each and every one of us?)

- **Honesty:** 'Would we be ashamed or would we maintain our dignity if all people suddenly gained access to our innermost thoughts and thus discovered exactly what was driving the decision?' (How would we react if we were subjected to a measure prompted by the same motivations?)

In summary, for any organisation the strategic implications of moving towards social sustainability will depend upon a number of factors. Such organisations need to think much more carefully about the consequences of **what they do, and how they do it**. They need a systems approach that analyses the impact of their behaviour on the capacity of others to lead a fulfilling life. They need to ask whether they would like to be subjected to the conditions they create, in political, economic, and environmental terms. They need to ask whether the processes they use are sufficiently participative, transparent, accountable and honest.

In his chapter it has only been possible to give an abbreviated account of the work that has been going on within TNS with regard to the social aspects of sustainability. Our main objective is to bring a better systems awareness to the debate about both Corporate Social Responsibility and public policy. As this work progresses we are building experience of working with organisations to put these ideas into practice. The concepts that underpin our analysis will continue to be tested through our network of science connections.

It is essential that social parameters are seen as intrinsic to the quest for sustainable development. They cannot be dealt with separately from the sustainability agenda. It is because we have become so good at pigeonholing problems that we so often fail to see the big picture. We need decision-makers who can link social implications to the wider ecological scene. System Condition 4 tells us about community relationships in a sustainable society in the same way that System Conditions One, Two and Three tell us about our relationship to natural resources. And if we are smart enough we will see that they are indivisible one from another.

Chapter 7

Putting the TNS Framework into Practice

Where and how can The Natural Step be used to help people make more sustainable decisions?

The System Conditions are tough. They have profound implications for the way we live our lives today. Whilst they are science-based, they are capable of practical application. The essence of TNS is applying scientific understanding to overcome divisions in society, and produce shared language and goals at the highest level. TNS is all about bridging between sectors, ideas, policies and societies.

In this section there are some examples of the way in which the TNS Framework can be applied. TNS covers a very wide and fluid range of issues, mixing research, education and strategic advice.

Over the last 15 years TNS has held a range of strategic dialogues: in Sweden they have focused on sustainable agriculture, forestry, transportation and energy. In the UK they dealt with incineration, renewable energy, sustainable sewage systems, bulk printing and the future of PVC. In the USA, they covered sustainable materials use, food, fish and fibre, and social change.

TNS has also:

- Held scientific consensus processes in Australia, South Africa and Canada
- Published over a hundred peer-reviewed papers
- Developed websites with case studies and research and education materials
- Educated tens of thousands of people in the TNS Framework
- Helped large corporations integrate sustainability into their strategy and activities

- Helped many major municipalities integrate sustainability into their planning
- Developed an international sustainability Masters Program 'Strategic Leadership for Sustainable Development'
- Held numerous multi-stakeholder dialogues on sustainability
- Advised the senior members of the EU Commission
- Advised the EU Environmental Technologies Action Plan Report to Ministers
- Advised the Nickel, Cement and PVC industries on sustainable development
- Established a community level sustainability initiative in Whistler, Canada
- Pioneered sustainability dialogues with leading organisations in Japan
- Worked with Cambridge University Centre for Sustainable Development on a new systems perspective linking sustainability and technology

The above list shows that the TNS Framework cannot be pinned down to any sector of society or type of activity. All in all, the practical application of sustainable development can be a very confusing scene. There are many tools and methods available to help organisations move towards practical application. The Natural Step does not exist in isolation from these tools and methods. It is worth looking briefly at how different approaches relate to each other.

Environmental Management Systems (EMS) are standards for implementing and monitoring environmental processes and responsibilities within an organisation. The most common of these is the International Standard ISO 14001. The European Union has also established a standard, the EMAS, which similarly provides a set of administrative routines for environmental stewardship and good practice within an organisation. These standards do not address the wider issues of sustainability, nor do they set out to give strategic guidance. A new standard known as SIGMA is being developed which will aim to widen the scope of EMSs and The Natural Step has been involved in the early work on that initiative.

Life-Cycle Assessment (LCA) is a well-established technique for evaluating the environmental impacts of a product. LCA aims to measure comprehensively all impacts throughout the life of a product, in manufacture, use and disposal. These impacts concentrate upon direct and indirect environmental effects. Experience suggests that there is little strategic content to the Life Cycle approach. LCA tends to be rather narrow, focussing only upon that which can be measured, and therefore lacks a sustainability perspective.

A number of concepts have been developed over recent years which concentrate upon the size of material flows within society. Factor 4, Factor 10 and Ecological Footprinting each address in their own specific and technical ways the need to reduce material flows. In a similar vein TNS also calls for such dematerialisation, particularly with regard to System Conditions 1, 2 and 3. Yet dematerialisation alone is insufficient. Some material flows need to be phased out altogether, since even minimal leakages will lead to accumulation in nature. Other flows, more suited to a sustainable society, may need to increase. Society also needs to substitute unsustainable materials with sustainable ones. Many aspects of sustainability cannot be presented in terms of dematerialisation at all. These concepts are certainly major contributions to sustainable development but they do not deal with the whole picture, particularly in terms of the social and strategic aspects of sustainability.

In the UK a set of principles has been devised by the sustainability NGO, Forum for the Future, known as the Five Capitals model. The metaphor of capital stocks is used in this model to describe natural capital (resources and services from nature), human capital (our talents and human potential), social capital (all the institutions and relationships that make up human society), manufactured capital (resources used in the economic system) and financial capital (money and financial investments). This model encourages stewardship of these stocks of capital in ways that are sustainable. Un-sustainability means exhausting the 'interest' which flows from these stocks, and running down the capital itself. Although there may be debate as to the extent to which manufactured and financial capital can be substituted one for another, safeguarding the flows and stocks of natural and human capital is really the bottom line of sustainability. Natural capital is the territory covered by TNS in System Conditions 1–3, and human/social capital in System Condition 4. Thinking of manufactured and financial resources, as in the Five Capitals model, can be a helpful way of evaluating where an organisation stands today with regard to sustainability, as step B of the TNS Framework (see Chapter Five).

Yet another concept, Natural Capitalism (NC), deals with the integration of societal flows of resources within the carrying capacity of the ecosystem. It uses the terminology of economics to address these issues. NC calls for recognition that economic growth will be limited by natural capital, that radical increases in resource productivity are essential, as well as a new valuing of natural resources within the market system.

Overall, the expansion of tools and concepts to deal with the complex issues of sustainability is encouraging. TNS continues to foster a science- and systems-based view of these issues which helps make sense of the many tools available. Without better-informed awareness, and a better systems perspective, many tools and concepts will fall short of their full potential. Making progress must mean encouraging more application of sustainable development in as many settings as possible. Success breeds success, and TNS works by finding and encouraging role models and champions to get on with applying the framework in their organisations. The examples that follow have been selected to show some different settings for the application of the TNS Framework. More stories can be found on the TNS website and there are several publications on the pioneering achievements of Swedish and other companies using the framework.

In compiling the examples I have covered a range of applications. From the business sector the stories include a practical construction project, corporate culture change, and sustainability learning in an international business. From the public sector I have included the story of TNS working with a local community, and sustainability workshops with a major Government agency. I also describe the science dialogues carried out by TNS, and give one example of the framework applied to a specific industrial material. In each case I have summarised key outcomes and key learning.

TNS in the business sector

The Natural Step has worked with people in some of the largest companies in the world. The examples of Electrolux, McDonalds, Ikea, Nike and Scandic Hotels have already been well documented. This work has ranged from short interventions to longer and deeper relationships. The relationship is always based upon our core values as a non-profit organisation. We try hard to learn from this advisory work, and to keep on enrolling the individuals we work with in the general process of sustainable development.

> *"TNS is a helpful, pragmatic approach which recognises the business realities without compromising the objective."*—Vivienne Cox, BP

Carillion: Pioneering sustainable construction

The major UK construction company **Carillion** (formerly Tarmac) was responsible for building and running a new hospital to serve the Swindon area, a £100m scheme. This was one of the first of the UK's new wave of health construction projects. The company had been already been talking to TNS about the strategic importance of sustainable development in the construction sector. Swindon was chosen as a Pathfinder Project to test out the TNS approach from the business perspective.

Construction is a tough business, where time and finance are critical. Yet it is also a sector which has massive environmental, economic and social impacts. TNS worked with the Carillion Project Team from 1999 through to the opening of the new Swindon hospital in 2002. Overcoming skepticism was the first hurdle; engineers, surveyors and builders were not used to this sort of approach. However, after in-depth awareness raising sessions, the Project Team began to apply the TNS Framework with professionalism and enthusiasm. They developed their own Sustainability Action Plan based on TNS System Conditions, and these were used to evaluate all materials and processes. The case study has been well publicised and is being used by the UK Government as the leading example of sustainable development as applied to construction.[8] A video about the project is available from Carillion.

8. DTI, Construction Best Practice Programme, Case Study 191, www.cbpp.org.uk; Carillion PLC, Wolverhampton, UK, www.carillion.com.uk.

Key outcomes:

- Reduced landfill—50% reduction of waste leaving the site compared to similar schemes
- Fewer construction vehicle movements—some 20,000 lorry journeys were eliminated
- Enhanced ecological impact—substituting naturally derived materials
- Zero-sum—balancing initial cost with consequent savings showed that sustainable construction need not be more expensive
- Can-do attitudes—employees began to see sustainable development as a core work objective
- According to Carillion's chief executive, "The hospital received top marks from the Building Research Establishment as one of the UK's most sustainable buildings, where more sustainable methods have generated identifiable savings of £1.8 million, including 30 percent less energy consumption and 35% less carbon dioxide emissions."

Key learning:

- Sustainability is a business opportunity
- Leadership in sustainability is critical
- Success in sustainable construction comes from the creativity and enthusiasm of individuals

Corporate cultural change

The Swindon Project tells an important practical tale of how the TNS Framework influenced a major development. It was a key part of a long-term engagement with a large and diverse company. It began with TNS's involvement with the construction and service company, Tarmac. TNS was first invited by a new Chief Executive to talk to the company's senior team as long ago as 1998. The management board divided between outright sceptics and those who thought that 'these people are telling us about something important, we need to listen and learn, and we can't do that on our own.'

TNS was offered the chance to work on two projects, the Swindon Hospital Project and a social housing scheme in the north of England.

The housing scheme was one of our first Pathfinder activities in the UK. It was based on a large, troublesome housing estate in Bradford, Yorkshire. Tarmac organised workshops for those involved and, with some encouragement from TNS, managed to get people to attend for whom that kind of experience was completely new. We worked with tenants' representatives, local politicians, bricklayers, as well as financiers and architects. It was an outstanding experience for us all. Major environmental improvements were put into the design.

But the outstanding result was on the social dimension, and our examination of System Condition 4 with that group. Amongst some of the innovations were: starting up a completely new, local cooperative to provide timber frames for this and other schemes; guaranteeing to train and employ the tenants on the project (most of whom were unemployed); and establishing special links with the local primary school. The company had predicted that everything that was not nailed down would be stolen or vandalised from day one. But our work helped assure that the estate became one of the safest and most crime-free sites the contractors had ever worked on.

The Swindon Project was very different, but TNS operated there on basically the same lines, raising awareness and understanding, and creating strength in the team through new and shared commitments to sustainability.

The strategic aim for TNS, and for senior colleagues in the business, was to build from these Pathfinder Projects enough learning, momentum and credibility to infect the wider Group. In the meantime, Tarmac Construction became the new Carillion plc. By that time the success of Swindon was becoming apparent, and TNS had enough of a reputation at senior levels to survive this upheaval. Awards and government accolades began to arrive from that scheme almost daily. A degree of internal doubt about all the good things at Swindon became apparent along some of the corridors at Carillion HQ. Here, Francis Meynell's findings take up the story.

> "Such were the stories emanating from Swindon and the TNS UK office, that members of Carillion Buildings Management Board sent Business Development Managers 'to challenge and really understand what was going on'. They facilitated three discussion groups and 'really listened'. While at first sceptical, these envoys became convinced that something special—a 'landmark building'—had indeed happened. One of them perceived that the people involved had started from very different positions. Some had been sceptical, while some had taken it on very easily. He also perceived that the whole team in Swindon had bought in, and had applied TNS's Framework to what they did in 'personally relevant ways'. In his view, the hospital project 'gave people the opportunity to innovate along environmentally and socially progressive lines that were beneficial to Carillion on a purely commercial basis'. Because of the good vibes from Swindon, TNS UK's Chief Executive was invited to give a two-day TNS workshop with Carillion Building's Management Board, which comprised one third of the plc business. The workshop was initiated 'to fill in the gaps, giving us the bigger picture'. It started with people asking the question 'Is it all too hard?', but ended with a sense of the practical possibilities to influence the situation positively. The Board 'pretty much bought into it'. The subsequent bi-annual operations conference—a very task-focussed pragmatic forum—was geared entirely to TNS and sustainability. A 'train the trainer' drive was devised within the organisation to 'infect' a representative sample of 10%—around 140 employees—of Carillion Building workforce with awareness of sustainability. An enabling Group was formed to facilitate and steer the initiative and involved three Board members."

At this stage of the TNS-Carillion relationship the joint goal was to infect the wider company, embedding sustainable development at the heart of its strategic operation. Those goals were helped greatly by a disposition toward change, to learning and to environmental and social values already embedded in the company culture. The then Chairman, Sir Neville Simms, was a prominent spokesperson for ethical business and better environmental performance by business in the UK. Nonetheless, getting strategic and application credibility for TNS methodology in his company, let alone this sector, was never going to be a pushover. The construction sector is a tough, commercial world. Most people working there have 'bullshit' detection capabilities as high as you will find anywhere.

Taking up the story again:

> "Resulting from its efforts to educate members of staff, reduce their ecological impacts, and build relationships with local communities and various stakeholders, (the business manager) saw the company as 'starting to become more joined up and to penetrate new fields'. He saw sustainability starting to migrate up to the plc level and across other parts of Carillion Group, for example, in the Infrastructure Management business. One of the project managers at Schal (another Carillion business) perceived the Swindon hospital project as beneficial for in-house training in his business unit, because it gave people 'real examples that they can latch on to "over and above" academic theory.' "

The Director of Engineering and Environment described himself as having learnt quite a lot since the company clarified its sustainable development trajectory. He had not noticed the issue of sustainability cropping up in conversations in the organisation eighteen months previously, whereas it had now become something 'commonly spoken about'. He attributed this to the 'general awareness raising that had happened, particularly through the Carillion Building and TNS programme.'

According to Carillion plc's Director of Engineering and Environment, TNS input accelerated the collaborative action and triggered enthusiasm. It helped Carillion to flag its sustainability credentials and win contracts. He has said that some of the project managers and designers had become evangelists for TNS. People were happier working within a common framework. The input of TNS had helped "fire the enthusiasm and people's belief in the possible, and helped coordinate action."

Key outcomes:

- Sustainable development is now one of the core values and business strategies of Carillion plc
- Real and practical examples of success from their own peers convinced enough Carillion people to get their commitment
- Sustainable Development is a winning business strategy

Key learning:

- Changing the culture of a complex and large organisation takes time and adequate investment; and success will be attributable to many factors
- Internal champions for sustainable development can make a real difference
- Top management buy-in as well as on-the-ground achievement are both essential
- Those promoting the change have to work with a light touch, always looking for opportunity to let people get on with it in their own way
- Organisations and the people in them change all the time; real success in culture change is only credible if it manages to 'hang-in-there' and survive that kind of turmoil.

Starbucks: An international opportunity for change

Every year, coffee lovers around the world drink 400 billion cups of coffee. As an environmentally and socially conscious company, **Starbucks** knows that a lot of resources go into the making of its products—from the coffee farms where most beans are grown, to the tons of paper cups, energy, and vast quantities of water needed to operate its retail stores. With more than 6,000 stores around the world and 20 million customers visiting a Starbucks coffeehouse each week, the company knows it has a prime opportunity to be a leader for sustainability within the industry.

In 2000, Starbucks began working with The Natural Step's sustainability framework and the TNS team in the US, to build on the company's long-standing commitment to the communities where it does business and the environment. Starbucks began using The Natural Step

Framework to assist in establishing sustainability focus areas and to develop tools to measure its ecological footprint.

Using The Natural Step's principles for sustainability as guidelines, Starbucks pulled together mid-level managers from its marketing, operations, finance, human resources and other departments to establish a system to measure and focus its environmental performance. This team worked with The Natural Step to map environmental and social impacts, including its inputs, outputs and material and energy flows across its network of suppliers, distributors and retail locations.

Key Outcomes:

- Coffee, tea and paper sourcing: Starbucks is tracking the number of pounds of shade, organic and fair trade coffee purchased, percentage of organic tea, and the post-consumer and unbleached content of its paper packaging
- Transportation: Starbucks is looking at minimising the environmental impacts of moving both people and partners (employees). Current focus is on providing alternative commuting options for their partners
- Store design and operations: Starbucks is working with the US Green Building Council on a new Leadership in Energy and Environmental Design (LEED) standard specifically for retail. In the interim, Starbucks is measuring the percentage of stores with recycling programmes, percentage of customers that use commuter mugs, and tracking its use of electricity, gas, and water per transaction-dollar of sales and square foot of retail space

Key Learning:

Step-by-step change: after two years, Starbucks has taken steps to steadily bring The Natural Step's Framework to life.

- Leadership: Starbucks commitment to sustainability has had top-down impact
- Integration: The company has provided training in The Natural Step Framework in order to continue building sustainability into its business model

Whistler: Putting sustainability into practice at the local level

Nestled at the foot of over 2,800 hectares of winter sports terrain that rises a mile above the valley, the resort community **Whistler** is nestled in the Coastal Mountains of British Columbia, Canada. Home to 10,000 permanent residents, Whistler is widely regarded as one of the best destination resorts in the world. Along with Vancouver, Whistler will co-host the 2010 Winter Olympics.

Whistler attracted the attention of Dr. Karl-Henrik Robèrt, who came in March 2000 for a vacation with his family. During his visit he addressed a number of Whistler audiences regarding The Natural Step and impressed business and political leaders with his clear and compelling explanation of sustainability. After one luncheon presentation, Chamber of Commerce members uncharacteristically lingered long after the presentation, captivated by the significance of what they had heard.

The timing of Dr. Robèrt's visit was ideal. Whistler had laid the groundwork for introducing sustainability to the community and the resultant Whistler Environmental Strategy, in particular, provided a comprehensive plan for environmental sustainability. But, there was no clear and compelling message to the community about the basic imperatives of sustainability. The Natural Step provided simple, common language necessary to communicate sustainability while also providing the necessary intellectual and scientific rigour.

Following Dr. Robèrt's visit, several local organisations, businesses and individuals formed a group called 'Whistler's Early Adopters', and began collaborating with TNS to create the 'Whistler, It's Our Nature' program, www.whistleritsournature.ca. They hosted a sustainability speaker series, built sustainability toolkits customised for households, businesses and schools, as well as a core resource toolkit with more comprehensive information, and participated in extensive training sessions for over 1000 staff members. The speakers series (with up to 1000 attendees at each event) was particularly effective, with its powerful messages about sustainability from renowned speakers. It generated what one business leader referred to as "talk on the street" about the opportunities associated with sustainability.

A Run of Success

Whistler has been integrating sustainability into resort operations for some time now, and more recently has become the first destination resort (and first municipality in Canada) to implement the TNS Framework in their overall approach to development. Consequently, their efforts have produced a number of successes that are beginning to move them towards sustainability, and in certain cases are producing financial gains as well. While many of these initiatives were conceived before Whistler started working with the TNS Framework, they have all been reinforced and propelled by its clarity and vision.

Some of these successes include:

- A transformed transit system that has grown over a eleven year period from five buses and 300,000 riders to a fleet of 24 buses and 2.6 million riders;
- An award-winning, pedestrian-oriented village surrounded by neighbourhood 'clusters'—allowing easy service by public transit and trail networks;
- Approximately 4,200 resident-only beds (1/3 of the total beds needed for Whistler employees) created through the establishment of the Whistler Housing Authority to address affordable housing issues in what has become an expensive real estate market;
- The Emerald Forest Conservation Project—a project to re-focus land development to a parking lot near the central Whistler Village, rather than private land, resulting in the conservation of 57 hectares within a much larger network of protected, natural areas;
- Financial returns to the community—excess tipping fees from Whistler's landfill raise $300,000 annually for an Environmental Legacy Fund. Interest on this rapidly growing fund goes toward financing local environmental projects;
- Pesticide-free parks—with biological, steam and mechanical weed and pest control only, and
- Geothermal heat exchange systems that heat and cool the Spruce Grove Community Building and the Beaver Flats resident housing project.

Many more of these examples continue to 'grow' organically out of the community dialogue and learning in combination with the stated commitment of senior community leaders. In their entirety, these accomplishments led the Federation of Canadian Municipalities (FCM) in 2002 to give Whistler a national award for 'best comprehensive sustainability initiative'.

Inspired in a large part by this new thinking around sustainability, the local government authority has prepared a 'Comprehensive Sustainability Plan (CSP)'—a very ambitious high-level plan based on TNS Framework and extensive community dialogues. Of course, there are debates around specific policy and project decisions. The CSP helps Whistler move strategically into the future with its eyes open to the walls and especially the opening of the funnel, as defined by TNS. This includes its continuing influence on the sustainability efforts regarding the 2010 Olympic Games. As one otherwise rather sceptical elected official remarked: "If you don't know that we use TNS to define sustainability in this town, then you have no business running for public office." This statement is an indication of the community-wide learning around sustainability that has taken place since Dr. Robèrt first visited Whistler in 2000.

Key outcomes:

- The CSP is probably the best example to date of comprehensively applying the TNS Framework to a local community
- It uses the TNS System Conditions to derive its own sustainability principles
- It has articulated a 2020 vision intended to guide all major planning and policy initiatives towards sustainability
- It has framed its plan by applying the A,B,C,D Method to its own situation
- It frames all specific strategy areas (e.g. energy, housing, economic development) within overall sustainability constraints
- It draws on TNS's new work on social sustainability—integrating self-organisation, diversity and interdependence—throughout the CSP's strategies on housing, affordability, social services, arts and culture

Key learning:

- The travel and tourism industry, expected to grow at 4% a year over the next two decades, depends upon inherently unsustainable technologies. Air traffic alone produces more and more carbon dioxide. Tourist roads being built through pristine areas can cause environmental degradation. Through its work with TNS, Whistler is finding ways of operating a thriving resort within a vibrant community whilst making real progress towards socio-ecological sustainability.

NHS: Sustainability awareness in a Government Agency

The National Health Service (NHS) Purchasing and Supply Agency (PASA) was established in April 2000 as an executive agency of the UK Department of Health. It is intended to take the lead role in modernising and improving the performance of purchasing and supply within the NHS, benefiting both patients and the public. The NHS is one of the largest organisations in the world, employing more that one million people and purchasing goods and services totalling over £15 billion each year.

Purchasing has direct social, environmental and economic impacts, and presents an ideal opportunity to reduce the risk of health inequalities that can come with a degraded environment, unemployment, poverty and social exclusion. Sustainable development is a key objective within the PASA corporate plan and business plan. Senior managers realised that they needed to do more to spread understanding within their organisation about sustainable development. Working with Forum for the Future and TNS, a sustainability workshop was developed for all staff to attend.

Key outcomes: The course brought out various elements of TNS Framework.

- 264 staff attended workshops
- very positive staff feedback achieved
- reduction in scepticism regarding sustainable development as a concept

Key Learning:

- enthusiasm from other parts of the NHS to run similar courses
- training needs to be motivational and non-threatening
- messages must be consistent with other activities happening concurrently, especially central government programmes
- people need to understand why they are being asked to change behaviour
- at work training can only be one part of a wider programme for sustainable development to be embedded

Science and dialogue

In addition to advising people in organisations, TNS has also built a track record in convening multi-stakeholder dialogues around key sustainability issues. Our experience has been that the science underpinning the TNS Framework can act as neutral ground. It provides a platform which people are prepared to step onto and share their views and experiences. It is possible to convene dialogue with organisations and individuals who would not normally get involved together. TNS as an organisation does not lobby or campaign on any particular cause. We are essentially non-judgmental. But we certainly know and care deeply about the prerequisites for sustainable development, with a unique science-based analysis that can be brought to any subject. We try very hard to be rigorous in that analysis, which means we do not necessarily arrive at easy or popular answers.

PVC: A sustainable material?

In a sustainable society, can there be a place for PVC? This might seem an odd topic for The Natural Step to be involved with. The TNS's involvement arose from work done in the UK several years ago with manufacturers and retailers. TNS was asked to take that work to a deeper analysis. PVC is a controversial material, and most commentaries seemed to be polarised between those who wanted a total ban on PVC and those who wanted to see much more of it. The TNS evaluation, published in July 2000, identified five major challenges. To make progress towards PVC becoming genuinely sustainable, the industry should make a long-term commitment to:

- becoming carbon neutral
- becoming a closed loop system of waste management
- ensuring that releases of persistent organic compounds from the whole life cycle do not cause systematic increases in concentration in nature
- reviewing the use of all additives consistent with attaining full sustainability, and especially committing to phasing out substances that can bio-accumulate, or where there is reasonable doubt regarding toxic effects
- raising awareness about sustainable development across the industry, and the inclusion of all participants in its achievement.

The last of these challenges has led to an initiative in the UK to establish a Stakeholder Forum for Sustainability. The initiative follows extensive work to identify all those organisations having an influence on the material. It is a very extensive and complex list. It is no good expecting progress to come only from the 'industry'. Users, retailers, regulators and Government, all have a role to play. Hopefully this initiative will produce a more objective debate and give an informed sustainability direction to further research and innovation. So far as the more technical challenges are concerned, some companies in the industry are working hard at more sustainable practices (notably Norsk Hydro and EVC in Europe).

Key outcomes:

- The identification of unambiguous sustainability tests for this important material—these are extremely tough tests but they represent the bottom-line for any hope of seeing this material endure the transition to sustainability
- Real and measurable progress by some of the manufacturers involved on emissions, additives and recycling—whilst others demonstrate very little change to practices that will never be acceptable from the sustainability perspective
- A positive initiative in the UK to bring together all stakeholders with a view to identifying the next steps in research—it is not just the manufacturers who are involved here; a complex range of stakeholders has to be engaged in the process

Key learning:

- PVC involves a complex chemical process and an equally complex web of users, suppliers, regulators, retailers, etc—it requires good systems thinking to get a clear understanding of the issues
- By identifying the systems obstacles to sustainability we have moved the debate to a new level, a level which is hopefully helping to focus on the very real and very large gap which the industry has to close, as opposed to investing time and energy in defending the indefensible.
- There can be a positive approach, solutions-oriented, to such a controversial product if those involved are prepared to invest time and energy in making progress.
- This material is so commonplace around the world and so widely used in everyday life (for beneficial purposes as well as for less essential uses) that its future management and as an issue cannot be left to chance nor to ill-founded judgements.

Chapter 8

22 Questions and Answers about the Science of the TNS Framework

This final chapter presents a series of questions and answers based upon a dialogue between Karl-Henrik Robèrt, Jamie MacDonald (TNS Canada) and Dave Waldron (Blekinge Institute of Technology). It is intended to enhance the understanding of both the science and the practical application of the framework.

1. Question: What does it mean that the TNS Framework is 'scientific'?

Answer: It is developed by scientists, and scrutinised in peer-review processes.

The framework is derived from a scientifically relevant world-view, and is elaborated within a relatively new field of science, systems thinking. This field covers issues such as system dynamics, chaos theory, game theory, resource theory, and planning in complex systems. Having been initially explored by physicists and meteorologists, 'systems thinking' has become truly multidisciplinary. The TNS Framework has been developed in an international consensus process amongst scientists from many different fields of expertise such as Physics, Chemistry, Ecology, Toxicology, Medicine and from social sciences such as Organisational Theory, Management and Economy. The framework and its applications are published in peer-reviewed scientific journals in the fields of Sustainable Development, Ecological Economics and Industrial Ecology.

2. Question: But isn't it true that scientists do not always achieve consensus?

Answer: That is true, but the TNS Framework is based on consensus.

The difference between the openness to new ideas in the scientific community, and that of the general public, is that the former is characterised by peer-review to create as unbiased knowledge as possible. This means that articles from various fields are scrutinised not only by the editor but also by reviewing scientists from the same field of expertise. Studies are then systematically checked against whether they are built on relevant scientific methods, whether the results are discussed in a logical way, and whether reference was made to previous knowledge in the respective field. This is different from a public discourse in which the expression of all points of view—even scientifically flawed arguments—are allowed for democratic reasons.

3. Question: Isn't it possible for even established scientists, publishing their findings in peer-reviewed journals, to be wrong?

Answer: That is true, in fact the scientific methodology with its peer-review process is about disregarding the 'authority' of individuals—it is only results and logic that count.

It is inherent in the scientific process to always question old and established knowledge. In fact, the whole scientific methodology revolves around self-criticism to reveal flaws in previous ideas and paradigms; otherwise the search for new knowledge would be built on flaws and be inefficient. The number of scientists behind new ideas is not what counts—only data and logical interpretation of data. It only takes one person to show that thousands of scientists may have been wrong. The TNS Framework has not been refuted, and in fact has grown stronger over the years through increasing scientific scrutiny.

4. Question: But isn't there yet another problem with scientists—they may have a political agenda too, like anybody else?

Answer: Yes, environmental scientists too can have political agendas.

Sadly, on occasion they may selectively cite data or otherwise attempt to illegitimately strengthen their cases. There is a difference, though. The scientific community is large and diverse, and the penalties for scientists doing bad science are severe. Scientists must subject their work to peer review if they are to maintain their reputations in the community, and their reputations are vital to getting the rewards that community has to offer. Scientists don't have to be right, but they've got to be honest, and the system is designed to keep them that way.

5. Question: Isn't it true that the scientists behind the TNS Framework are not following the scientific mainstream?

Answer: Studying how scientific knowledge can be better used for decision-making is a potentially dangerous field of expertise for the individual scientist, yet it cannot responsibly be handed over to any other community than the scientific.

The TNS Framework is developed in the new scientific arena of Systems Thinking, and a branch of its own in this arena—'Planning in complex systems'. It's about scientists taking responsibility for how scientific data is used for decision making in society. This is a relatively unpopulated scientific arena partly because it is not without risks for the individual scientists active in it. There are risks involved for the individual scientist's integrity, for instance, by being taken hostage in a political process. Cornerstones of good science are to: (1) avoid all types of bias, (2) be thorough when it comes to restrictions and system boundaries, and (3) allow the time it takes to penetrate complex areas with scientific methodologies. These aspects run in the face of using science as an argument for this or that standpoint. This is particularly relevant, since societal decisions and policy-making cannot be directly arrived at from scientific knowledge—such attempts are called 'the naturalistic fallacy'.

There is no way we can avoid filtering scientific knowledge through our value systems for the choices we make.

However, the fact that science cannot directly inform policy making is in no way an excuse for the scientific community to resign from its responsibilities as regards societal policy making. Who has the responsibility if scientific data are misunderstood and/or sub-optimised for decision-making? The scientists behind the TNS Framework seek generic principles for decision making as such. Rather than seeking more data and taking stands as regards their respective importance for decision-making, we wanted to find generic principles for anybody to structure any kind of data in a comprehensive way for decision-making. Such principles should help decision-makers ask the right questions, and discover which data are relevant for decisions, which data are not, and which data are lacking. Thereafter, the individual's personal value system can be allowed to enter the picture. In this way we can elaborate together—in dialogue—the basic standpoint, avoiding the need to argue. If successful, this will increase the chances of making informed decisions, reduce the risk of misunderstanding, and help us see true differences of opinion much more clearly. Without knowing how to structure knowledge, we all tend to seek knowledge, but to drown in information. The overall methodology to find such generic layers of knowledge was to simply turn the traditional question 'What do we disagree on' to 'What can we agree on?' The experience was rewarding upfront and throughout the process: scientists can reach consensus at the basic principle level, but consensus ends somewhere on the way to higher levels of principles and details, where complexity inherently escalates dramatically.

6. Question: Is consensus always something to strive for—aren't polarities part of life?

Answer: It depends on the systems level—consensus is an effective methodology when it comes to the identification and phrasing of basic principles, but may be of little or no value when it comes to choosing between the specific solutions that comply with those basic principles.

Consensus, i.e. when everyone agrees with everyone else, is not necessarily a positive thing. Many people are highly sceptical of consensus, rightly fearing that it can lead to the formation of sectarian groups. Alternatively, consensus can lead to 'adjusted' results because a group that has to reach consensus will often avoid contentious issues. Other common standpoints are that 'polarity is good for creativity' and 'freedom of thought is more important than anything else.' Consensus has a controversial reputation because it has often been attempted at the wrong levels of systems.

Consensus used at the appropriate level in a system, i.e. at the generic 'trunk and branch level', is essential for effective collaboration. Many people can help each other define such principles through dialogue, and scrutinise claimed generic principles with greater accuracy than any individual on his or her own. This is not only intellectually rewarding for comprehension, it is also a way of leadership. To be effective as a shared mental model for cooperation, the principles must be agreeable with regard to the values within the group, commonly applicable, and described in a way that makes sense to all participants in the group. For a group with diverse values (i.e. society), the principles are also, preferably, neutral.

Once consensus has been reached at the basic level, it is usually well worth taking the lid off and allowing individuals to freely draw the conclusions of the basic principles. Often, creative individuals in the group come up with smart and concrete solutions that benefit everyone else. In other words, one looks for consensus at the level of 'trunk and branches', and free creativity at the 'foliage' level.

7. Question: Isn't the TNS Framework an example of over-belief in the role of science?

Answer: Distinction must be made between the model as such, and the art of using it.

The TNS Framework itself is designed to be a shared mental model for cooperation around sustainable development that is generic enough to be applied to any activity at any scale. To make that possible, we have utilised scientific thinking, methodologies and scrutiny to the best of our knowledge. Applying the framework, on the other hand, is where the art begins. It's about community building, genuine creativity, ethics, group dynamics, common sense and psychology. It's the musicians and chess-players mastering the basics who can improvise.

8. Question: Isn't the TNS Framework an example of a deterministic model?

Answer: The 'endpoint' of the TNS Framework is compliance with basic principles that allows for continuation of biological and cultural development.

A deterministic model is one that builds on the idea of a specific endpoint that is often interpreted as inevitability or helplessness in the context of an ultimate fate. The TNS Framework aims at compliance with the constraints of sustainability (the System Conditions) that are prerequisites for a continuation of biological and cultural evolution. That's how sustainable development is defined—first (i) a development towards sustainability, then (ii) continuous development within sustainability constraints. The System Conditions do not specify a deterministic end, only principles by which whatever state we choose must be scrutinised. In fact, today we are living a deterministic model heading towards an end to biological and cultural development. The framework is there to help us turn our paradigm into a non-deterministic one.

9. Question: Isn't the TNS Framework superficial/simplified?

Answer: A structured bird's eye perspective avoids reductionism and is not an alternative to more detailed levels of knowledge; on the contrary, it makes better use of such.

The TNS Framework represents a framework that is comprised of basic principles for the structuring of Sustainable Development at the largest systems level: 'Individual, within Organisation, within Society, within Biosphere'. It is only after a thorough structuring of the system that we can avoid reductionism and understand its details in a meaningful way. Adopting a bird's-eye perspective makes it simpler to be a successful planner in a complex system, but it doesn't mean 'simplified' in the sense of reducing, i.e. neglecting, any of the complexity of the system. On the contrary, an effective use of the bird's-eye perspective is naturally followed up by higher degrees of resolution and details—'Simplicity without Reduction'. That's how we can avoid a superficial and/or confusing interpretation of details. It is out of respect for complexity in complex systems that we need a thoroughly structured overall framework that is robust enough to carry the weight of all relevant details. To understand the basic principles of chess is not a simplified way of looking at chess, but a prerequisite to understanding the higher degrees of the game's sophistication.

10. Question: But does the TNS Framework really cover everything?

Answer: In line with the last question and answer, distinction must be made between what the framework covers, and what it contains.

First of all the framework is not a rule of thumb for how to live our lives, so it really doesn't cover everything. It is a framework for the essentials of planning for a sustainable society in the ecosphere and doesn't—for instance—cover the ethics of how to manage domestic animals (since it is theoretically possible to treat domestic animals badly in a sustainable society). Secondly, as a framework of basic principles for sustainable development (SD), it really covers SD in the sense that it can be used to structure all data, measures and tools that are relevant for SD. But the framework doesn't contain these elements per se.

11. Question: But is it really realistic to foresee a future without (I) Mining, (II) Chemicals, and abuse of (III) Ecosystems and (IV) People?

Answer: The System Conditions do not state such things.

All those things can occur within a sustainable society so long as leftovers do not continue to increase more and more in concentration in the ecosphere (I, II), nature is not physically encroached on more and more (III), and the obstacles to people's ability to meet their needs do not increase more and more (IV). Thus, from a theoretical point of view we could use (I) mining as well as mined materials in the sustainable society: all the way until such fuel sources are exhausted and new energy systems need to be put in place—as long as the waste from such use is deposited in a way that doesn't allow systematic increases of CO_2 or other leftovers in the ecosphere. Likewise, we could use (II) all kinds of chemicals in the sustainable society. From a theoretical point of view we can even imagine using compounds that are persistent and foreign to nature, as long as they are prevented from leaking into natural systems. In a sustainable future we can even imagine a (III) road built on fertile land and (IV) political corruption—as long as these do not systematically increase. The System Conditions denote constraints for the survival of civilisation, they do not describe utopia.

12. Question: But isn't this a mechanistic definition of 'sustainability'? Where is, for instance, 'Spirituality as part of our Nature' or 'Conservation as expressed in Deep Ecology'?

Answer: A structured bird's eye perspective of a system can help us remember, and make better use of, all relevant aspects of the system.

It is true that 'Nature' has a value in itself that goes beyond providing resources and services to mankind. In fact, it was such drivers that triggered the foundation of TNS. From a systems perspective, it is a flaw to believe that there is a clash between respect for Nature on the one hand, and respect for rational human progress on the other. We are part of Nature ourselves. It is not likely that humans will ever comply with the System Conditions without a deeply felt respect for the inherent values of nature (System Conditions 1-3) as well as humankind (System Condition 4) including the spiritual aspects of those things. The framework is comprised of basic principles, as overriding 'checkpoints' of future sustainability. Thereafter we must all, as individuals, manage our transition towards these checkpoints by collecting the knowledge we feel relevant for our way of being and acting. Religion as well as 'Deep Ecology', for instance, can certainly supply valuable knowledge and experiences. And conversely, wanting to safeguard the system we are part of can be seen as a prerequisite for the higher values we seek.

13. Question: But mustn't sustainability be defined also in economic terms?

Answer: Social and ecological sustainability are goals and our economy can be made into one of many means to reach these goals.

The System Conditions define the goal of sustainability at its two fundamental levels—ecological and social sustainability. Our way of managing our economy can be made into a means of arriving at those goals. The basic definition of 'economy' is to use resources in a way that is efficient with regard to our goals, i.e., 'economy' represents one of many means to arrive at any goal.

Means and goals should never be confused with each other. For SD this is of particular importance. First of all, being economically powerful is neutral to 'good' or 'bad'. In fact, it is some aspects of our current society's industrial economy, and the way we measure the strength of it, that provide the largest threats to social and ecological sustainability on the global level. Secondly, we can be patient and wait for compliance with our goals, but we cannot wait for economic sustainability. If we want to arrive at any goal, good or bad, we need to be economically sustainable now and throughout the whole process. For these reasons, the TNS Framework deals with economic sustainability as a strategic means of arriving at ecological and social sustainability, and does so under 'D' in the A,B,C,D methodology. The TNS Framework provides principles for prioritising and managing actions that are consistent with this economic means to sustainability. Managers who master the art of this decision-making are key to successful application of the framework.

14. Question: Are the TNS System Conditions really laws of nature? (Isn't this an example of the Naturalistic Fallacy?)

Answer: The TNS System Conditions are basic principles for sustainability, and they are derived from the laws of nature.

A law of nature, such as the gravity law, cannot be violated. The System Conditions are basic principles, i.e. conditions for sustainability. Unlike laws of nature, you can violate principles and experience the consequences. However, this difference does not mean that 'principles' are inherently weaker than 'laws of nature' from a scientific/intellectual point of view. Principles, such as the System Conditions, may be derived from laws of nature in a way that is intellectually rigorous with regard to what the principles are for. The System Conditions are scientifically valid principles for sustainability, like the principle of breathing is a condition for survival of animals and humans. These things follow from laws of nature, such as the laws of thermodynamics, but it is not a law of nature to design sustainable societies or to keep breathing.

15. Question: Is the fourth System Condition as scientific as the other three?

Answer: The fourth System Condition is as logical and relevant for sustainability as the other three.

In fact, the whole endeavour of SD is about meeting human needs within ecological constraints. What this question probably implies is that it is often more difficult, from a scientific perspective, to assess System Condition 4 than System Conditions 1–3. It's more difficult to know if violations of human rights are increasing or decreasing than it is to assess whether mercury is increasing in biota or not. However, whether something is easy to measure or not is not directly linked to its accuracy or to its applicability. Looking both ways before we cross the road is a solid principle for staying alive in traffic, though it is extremely difficult to assess this principle by scientific means. The principle follows from logic and is easy to apply, but has not been verified by scientific means.

16. Question: Is the TNS Framework better than other models?

Answer: An overall framework is not an alternative to various concepts and tools.

There are many good models, tools and concepts for SD such as the Hanover principles, Natural Capitalism, Cleaner Production, Zero Emissions, Ecological Footprinting, Fritjof Capra's eco-literacy principles, etc. The Natural Step was not developed to replace any of these. It was developed as a unique overall framework for SD. Since it embraces the whole picture by means of strictly derived basic principles, it can be used for—amongst other things—demonstrating the relationship between other concepts, selecting and applying them for various purposes, and making better use of them.

17. Question: What is so great about the System Conditions? Aren't there other sets of principles around?

Answer: The System Conditions are the only principles that are designed for Backcasting from Principles, and only make sense in that context.

The intellectual breakthrough behind the TNS Framework is not so much the System Conditions, though this is often the perceived impression of TNS. The System Conditions were preceded by a logical sequence of reasoning: (i) To be strategic in a system we need to understand (ii) enough of the system (mainly the basics) in order to be able to have a (iii) accurate definition of what we want within the system.

However, nobody can look into the future. But we can invent it, as Einstein said. This means we define it in one way or another and then go for it, and this is referred to as 'backcasting'. There are two ways of doing this—planning from scenarios (i.e., the 'jig-saw' analogy) and planning from basic principles of success (i.e., the 'chess' analogy). The Natural Step has pioneered the latter. There are many good principles for SD. However, the System Conditions are the only principles that are designed for backcasting. To be robust for backcasting, such basic principles need to be: (i) derived from a scientifically relevant world-view, (ii) necessary for sustainability, (iii) enough for sustainability, (iv) general enough to be applied to all kinds of activities and scales, (v) concrete enough to guide our thinking and (vi) non-overlapping, to allow comprehension and development of indicators. Without this context, an overemphasis on the System Conditions is counterproductive, in particular if it is done in competition with other principles or concepts for SD.

18. Question: Isn't the framework too theoretical—can it really be applied in reality?

Answer: With the framework follows a formalised and easy-to-manage manual for concrete analyses and actions—the A,B,C,D methodology.

The A,B,C,D methodology is a concrete manual for how to facilitate constructive dialogue about very concrete changes on all levels. Once basic principles are understood, it gets easier and easier to apply them in reality, and vice versa. It may seem like a shortcut to skip a thorough basic understanding and dash into action, but in the long run this is more difficult.

19. Question: But isn't it very difficult to apply the framework in reality?

Answer: To be a good player in any complex system is difficult, but it is easier if we have a thorough basic understanding of the system.

It's like chess. Understanding the rules of chess may take some time, but the difficult part is becoming a good player, and requires training and experience. It goes without saying that it is easier to learn the basic rules of chess than to blindly apply 'trial and error', or 'simplicity without reduction'.

20. Question: Shouldn't we have all kinds of other tools and concepts as well?

Answer: A well-structured overview is not an alternative to tools; on the contrary, it makes better use of them.

The TNS Framework is not one of many 'tools' or 'concepts' such as LCA, ISO 14001, and Factor Ten. It is an overriding framework, and should, as such, not only be used for comprehension of the overall picture and design of our overall goals but also for selection and design of the tools and concepts we need for the transition. Different tools are appropriate for different endeavours, and the tools we eventually chose should be designed and used in a way that complies with the overall picture.

21. Question: But if we want to really change society, is it feasible to perceive business as the main force behind such a great ethical endeavour?

Answer: Business is not the main force, just one of many crucial players that need to engage in dialogue with the other crucial players, so we need a generic framework that can be used by all.

Business is the economic engine of society. And business carries one of the largest responsibilities for our current unsustainable societal course. However, business represents but one of many important actors, others being the general public, various kinds of institutions such as universities and the United Nations, and governments on the national as well as municipal levels. Having a shared mental model for SD, based on generic principles, is helpful in facilitating meaningful synergies across all societal sectors. An example is when proactive business corporations abandoned organised lobbying for fossil fuels and instead turned to politicians, speaking in favour of higher taxes on obsolete and less sustainable practices such as using fossil fuels. A rule of thumb could be to allow as much as possible to occur from shared understanding and free will, and then manage the rest, i.e. what remains to be done for the common good, through government and legislation.

22. Question: Finally, if you were to express, in just a couple of sentences, the unique quality of the TNS Framework, what would that be?

Answer: The framework consciously applies a new type of system boundaries—basic principles of success.

Whenever we want to plan ahead in a complex system we need to analyze the situation. The first question often is—what are the system boundaries for our analysis? The larger the complexity, the stronger is the temptation to apply very tight system boundaries. But that inherently leads to reductionism, and for sustainability you cannot apply regional system boundaries—it's the whole world that counts. Nor can you apply system boundaries defined within a single field of expertise such as toxicology or economy—most fields of expertise can be made operational, for the survival of civilisation. So why not apply basic principles of success for whatever you want to do? It may seem a bold statement, but who can deny the rationale for letting everything in the whole system—regardless of locality or field of expertise—that is essential for success, be taken into account?

Sustainable Solutions: a systems approach

the NATURAL STEP

"In a sustainable society people are not subject to conditions that systematically undermine their capacity to meet their needs."—TNS

Systems approach—the only way to see the connections, the only way to avoid blind alleys, to achieve a shared understanding and common agenda—there is no alternative.

We will not take care of nature until we learn to take care of each other.

We are all part of the same system—good leadership amidst the complexity of that system demands that we understand it.

The Natural Step 2004

Bibliography

Chapter One

Robèrt, Karl-Henrik, *The Natural Step Story: seeding a quiet revolution*, New Society Publishers, Gabriola Island, 2002

Chapter Two

Daly, Herman. *Beyond Growth: the economics of sustainable development*, Beacon Press, Boston, 1996

Gore, Albert, *Earth in the Balance*, Earthscan Publications, London, 1992

Ekins, Paul, *Economic Growth and Environmental Sustainability*, Routledge, London, 2000

Margulis, Lynn, *The Symbiotic Planet: a new look at evolution*, London, 1998

Gladwin, T.N. 'Why is the Northern Elite Mind Biased against Community, the Environment and a Sustainable Future?', in *Environment, Ethics and Behaviour*, New Lexington Press, San Francisco, 1997

Chapter Three

Porritt, Jonathon, *Playing Safe: science and the environment*, Thames and Hudson, London, 2000

Bruges, James, *The Little Earth Book*, Alistair Sawday Publishing, Bristol, 2000

Hawken, Paul, *The Ecology of Commerce*, Harper Business Press, New York, 1993

Daly, Gretchen, 'Understanding Nature's Services to Societies', in Waage, S., ed., *Ants, Galileo and Gandhi*, Greenleaf Publishing, 2003

Azar et al, *Socioecological Indicators for a Sustainable Society*, Chalmers University, Gothenburg

Chapter Four

Holmberg, J., And Robèrt, Karl-Henrik, 2000, 'Backcasting from Non-Overlapping Sustainability Principles: a framework for strategic planning', *International Journal of Sustainable Development and World Ecology*, volume 7, pp. 1-18.

Chapter Six

Wilson, E.O., *Consilience,* Abacus Press, London, 1998
Anders and Bromm, *Profit Without Measure,* Nicholas Brealey Publishing, London, 2000
Capra, Fritjof, *The Hidden Connection: a science for sustainable living,* Harper Collins, London, 2002
Williams, G.C., *Plan and Purpose in Nature,* Phoenix Press, London, 1996
Zohar, Dana, *Rewiring the Corporate Brain: using new science to rethink how we structure and lead organisations,* Berrett-Koehler Publishers, San Francisco, 1997
Meadows, Donella, *Places to Intervene in a System,* Whole Earth, 1997
Weinberg, R., *One Renegade Cell,* Phoenix Press, London, 1998
Stiglitz, Josef, *Globalization and its Discontents,* Penguin, London, 2002
Maturana and Varela, *The Tree of Knowledge,* Shambhala, Boston, 1987
Bateson, Gregory, *Mind and Nature,* Hampton Press, 2002
Gray, J., *Straw Dogs: thoughts on humans and other animals,* Granta, London, 2002
Roszak, Theodore, *The Voice of the Earth: an exploration of eco psychology,* New York, 1982
Bentley, Tom, ed., *The Moral Universe,* Demos Collection, London, 2001

Chapter Seven

Robèrt, Karl-Henrik, Schmidt-Bleek et al, 'Strategic Sustainable Development: selection, design and synergies of applied tools', *Journal of Cleaner Production* 10, 2002
Rowland, E., and Sheldon, C., *The Natural Step and ISO 14001,* The Natural Step and British Standards Institute, 1999
Andersson, K.; et al, 'The Feasibility of Including Sustainability in LCA for Product Development', *Journal of Cleaner Production* 6, 1998
Hawken, Paul, Lovins, Amory, and Lovins, Hunter, *Natural Capitalism: creating the next industrial revolution,* Rocky Mountain Institute, 1999
Forum for the Future, *The Five Capitals Model,* Forum for the Future, London, 1999
Meynell, Francis, *Awakening Giants,* Open University, Milton Keynes, UK, 2004

Contacts

Australia
The Natural Step Environmental Institute Australia
Matt Green
Level 1, 132 Leicester Street
Carlton, Victoria 3053
Australia
tel: + 61 3 9653 6440
fax: + 61 3 9011 6124
austadmin@naturalstep.org
www.au.naturalstep.org

Brazil
The Natural Step, Brazil
Simone Ramounoulou
Willis Harman House
Rua Lisboa, 328 – Pinheiros
São Paulo – SP 05413-000
Brazil
tel/fax : + 55 11 3064 4630
contato@willisharmanhouse.com.br
www.willisharmanhouse.com.br

Canada
The Natural Step, Canada
Saralyn Hodgkin
43 Eccles St., 2nd Floor
Ottawa, Ontario K1R 6S3
Canada
tel: + 1 (613) 748 3001
fax: + 1 (613) 748 3372
shodgkin@naturalstep.ca
www.naturalstep.ca

France
Caroline Gervais
cgervais28@ifrance.com

Israel
The Natural Step, Israel
Mrs. Limor Alouf
7 Halimon Street, Ramot Hashavim
PO Box 3380
Israel 45930
tel/fax: + 972 97482492
ecorrect@netvision.net.il

United Kingdom
The Natural Step, UK
Vanessa Mamo-Mason
Overseas House
19-23 Ironmonger Row
London ECIV3QN
United Kingdom
tel: + 44 20 7324 3688
fax: + 44 20 7324 3635
info@naturalstep.org.uk
www.naturalstep.org.uk

Italy
Eric Ezechieli
The Natural Step
Via Tadino 26A
Milano, 20124, Italy
tel: + 39 02 2040 4728
cell: + 39 348 60 600 60
tnsitaly@yahoo.it

Japan
The Natural Step International, Japan
Sachiko Takami
C/o Scandinavian Tourist Board
Izumikan Gobancho 4F, 12-11
Gobancho, Chiyoda-ku
Tokyo 102-0076, Japan
tel: + 81 3 5212 1528
fax: + 81 3 5212 1122
takami@sun.interq.or.jp
website: coming soon!

New Zealand
The Natural Step, New Zealand
Rhys Taylor
National Coordinator
PO Box 69
Lincoln 8152
New Zealand
tel: + 64 3 325 6711
fax: + 64 3 325 2418
natstep@naturalstep.org.nz
www.naturalstep.org.nz

South Africa
The Natural Step, South Africa
Stephen Jacobs
The Learning Centre
Dreyersdal Farm
Bergvliet 7945, South Africa
tel: + 27 21 715 0526
fax: + 27 21 715 0325
sjakes@iafrica.com

Sweden
The Natural Step, Sweden
Mats Nyström
Det Naturliga Steget
Garvargatan 9C, 11221 Stockholm
Sweden
tel:+ 46 8789 2900
fax: +46 8 789 29 39
info@detnaturligasteget.se
www.detnaturligasteget.se

United States
The Natural Step, US
Leah Laxamana
116 New Montgomery Street, Suite 800
San Francisco, CA 94105, USA
tel: + 1 415 318 8170
fax: + 1 415 974 0474
services@naturalstep.org
www.naturalstep.org

TNS International
Garvargatan 9C,
11221 Stockholm Sweden
info@thenaturalstep.org
0046 8789 2900
Karl-Henrik Robèrt: Chairman
David Cook: Chief Executive and R&D
Magnus Huss: Development and Policy
Catherine Gray: Fundraising and
Communications (San Francisco office)
tel: + 1 415 318 8170

Sustainability Masters Programme
David Waldron
Project Leader,
Strategic Leadership Towards
Sustainability
Blekinge Institute of Technology
371 79 Karlskrona,
Sweden
tel: + 46 455 385522
fax: + 46 455 385507
sustainabilitymasters@bth.se
www.bth.se/tmslm